Birding Guide to West Virginia

by

The Brooks Bird Club of West Virginia

Compiled and edited by James W. Bullard

1999

Published and distributed by
The Brooks Bird Club
of West Virginia

P.O. Box 4077
Wheeling, WV 26003

Printed in the United States of America
First printing

Cover design and illustration of Canada Warbler in Mountain Laurel by Julie Zickafoose.

To Kyle and Eleanor Bush
who brought me into
The Brooks Bird Club

Preface

A few years ago, I learned that there were two states that did not have birding guides. This fact was presented to the members of The Brooks Bird Club of West Virginia at the annual meeting at Cedar Lakes in 1993. The membership unanimously passed a resolution to have the club create a Birding Guide to West Virginia since the BBC had been completing scientific studies involving the flora and fauna of West Virginia, especially birds, since the 1940s. These studies included Breeding Bird Surveys, Singing Male Censuses, and surveys of the flora, mammals, reptiles, amphibians in most of the counties in West Virginia and, of course, have been of great help in formulating this guide.

I wish to thank the more than fifty people who have contributed to the guide. They are listed in the back of the book with their telephone numbers since they have graciously agreed to receive calls to answer questions regarding current birding conditions.

I particularly wish to thank Arline Thorn who undertook to complete the layout, Greg Eddy and Jay Buckelew who did the technical editing, Julie Zickafoose who did the art work for the cover and Bill Thompson who provided guidance regarding the publishing of the guide. Finally, I deeply appreciate the encouragement, support and good counsel provided by my wife, Beth.

—James W. Bullard

Table of Contents

Part One

Introduction to Birding in West Virginia

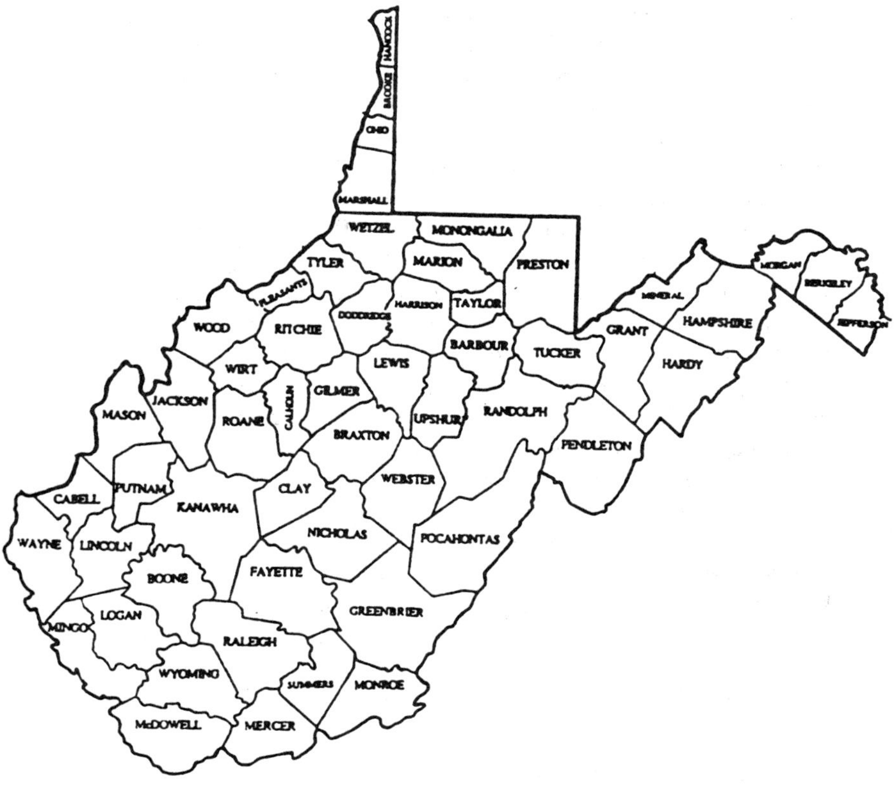

County Map of West Virginia

How to Use This Guide

The purpose of this guide is to help its readers find birds and plants in West Virginia. The guide covers the entire state which is divided into habitat-topographical regions. Near the back of the guide, you will find a check list of birds by habitat, status and abundance. **Birds of Particular Interest** describes those birds that are special to the state. By checking the index, one can locate where these birds may be usually found. The better **Birding and Botanizing Hot Spots** are described in detail by region , with an introduction and usually a map. There are many areas that offer excellent birding that are not mentioned in this book so bird and botanize as you drive along. When visiting any site, stop at the visitor's center or ranger station and pick up maps and bird lists when available.

In regard to the descriptions and maps, the roads are given by name or by letters: for example, Interstates as in I–77, federal routes as in US 50, state routes as in SR 55, county routes such as CR 28, and Forest Service Roads as in FSR 114. If you are traveling in a SW or NE direction, you can usually make good time, but if you are traveling in a SE or NW direction, expect to take much longer since the roads across the mountains have many switchbacks and are narrow, especially the county routes and forest state roads. Many of the directions are followed by a mileage figure (for example, 3.5 miles) which indicates the mileage from the last spot mentioned. The West Virginia State Road Map would be of great help in conjunction with this guide.

There is a list of volunteers in Part Three who have birded and botanized in West Virginia for years and who have participated in creating this guide. They are willing to accept calls to answer questions and offer current information.

The Brooks Bird Club Hot Line is 1 (304) 736-3086.

Visitor Information

Weather —West Virginia, due to its topography, has considerable variation in temperature and precipitation. The average reported temperature in spring is 52°, in summer 71°, in fall 54° and in winter 33° degrees. The Allegheny Mountains and Upland Section can be 10 to 15 degrees lower than in the Western Hill Section and in the Eastern Ridge and Valley Section. Rainfall is plentiful in all parts of the state, being heaviest on the western slopes of the high mountains of Randolph, Webster, and Nicholas counties, and lowest east of the Alleghenies in Grant, Hardy, and Pendleton counties.

Pests —West Virginia has a host of bothersome pests — snakes, ticks, horse flies, mosquitoes, no–see–ums and chiggers. Although snakes are not that common, be careful when hiking. Chiggers are a problem during the warm summer months in grassy and weedy areas since one does not realize that one has been bitten until the next day. They raise a welt which can itch for weeks. A small bottle of camphor from the drug store can alleviate the itching. Ticks and chiggers can be dealt with by tucking your pants legs into your socks and spraying before going into the field.

When to visit — The best time to bird and botanize West Virginia is during the spring, early summer and fall neotropical migration. There is an abundance of spring, summer, and fall flowering wildflowers throughout the state.

Where to stay — Stop at the Information Centers (see Part Three) and pick up the state road map, the West Virginia Lodging Directory, the Bed & Breakfast Directory, and the Camping Directory. All hotels, motels, inns and state park lodges listed with the West Virginia Division of Tourism and Parks may be contacted for room rates, facility descriptions, addresses and telephone numbers free of charge through the welcome/information centers. For more information call 1–800–CALL–WVA.

What to bring and wear — It is important to bring, at a minimum, your favorite bird and botany guides, binoculars and a birding scope, a hand lens, a hat, rain gear, and a warm (down) jacket, for it can be quite cold in the early morning even during the summer.

Short History of West Virginia

West Virginia, the 35th state to join the Union, gained statehood on June 20, 1863 at the time of the Civil War. Each of the bordering counties along the line between West Virginia and Virginia was given the opportunity to decide which state that county would join. The 1994 census showed a population of 1,822,000 that lived within the 24, 282.45 square mile area of the state. The highest elevation of 4,861 feet is at Spruce Knob in Pendleton County and the lowest elevation is 248 feet at Harpers Ferry in Jefferson County. There are 55 counties in the state.

Vegetation and Habitat

Eleanor Bush

"West Virginia is becoming a wilderness." J. Lawrence Smith made this statement during a lecture that he gave to the Brooks Bird Club of West Virginia at a Jackson's Mill meeting on February 23, 1996. Yes, we need but note that agricultural land is being abandoned so reversion to native and non–native species is occurring. Strip–mining laws are resulting in reclamation, and the soil has improved, allowing revegetation to take place. Pollution laws are benefiting rivers and streams so aquatic vegetation is being reestablished. Population has decreased due to lack of industry so that human interference with natural areas is declining. With less fragmentation of forests, ecologists, conservationists, and environmentalists may well welcome a new period of less destruction. More than ever it is timely and necessary that a guide to West Virginia birds and their habitats be published.

In *The Flora of West Virginia* (2nd ed., 1977), Dr. Strausbaugh and Dr. Core divided West Virginia into several types of forest vegetation. West of the Allegheny Mountains is the Western Hill Section, composed of the Central Hardwood Forests; centrally, the Allegheny Mountain and

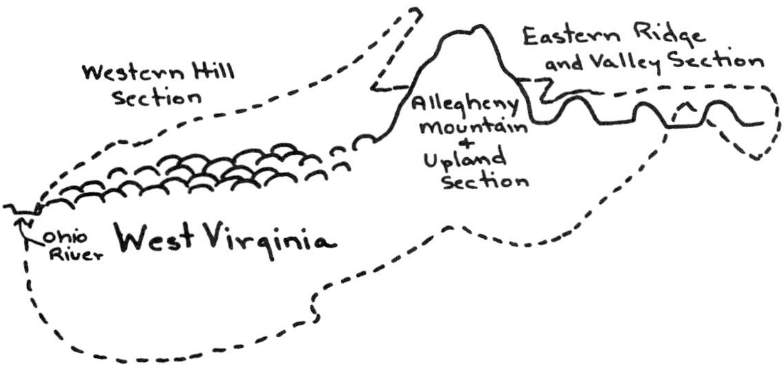

Drawing showing the three phytogeographic regions on an east–west profile map of West Virginia.

Upland Section includes the Northern Forests; and the Eastern Ridge and Valley Section includes Oak–Pine Forests. Within these sections there is diversity as well as overlapping of plant species.

The Western Hill Section may be described as a mature plateau (Allegheny Plateau) which is between the Ohio River and the mountains. It is part of the Central Hardwoods, composed mainly of Oak–Pine areas and Cove Hardwoods. Trees such as white oak, basswood, white ash and other deciduous forest trees are dominant. Shrubs such as dogwood, elderberry, witchhazel, hazelnut, wild hydrangea, azalea, and spice bush are common, along with spring wildflowers such as violets, bloodroot, spring beauty, plume lily, may apple, trillium, and Dutchman's breeches.

The Allegheny Mountain and Upland Section includes the highest elevations. Vegetation belongs to the Northern Forests and is composed of the Northern Evergreen Forest and the Northern Hardwood cover types. Outstanding here is the Northern Evergreen Forest of red spruce which is an extension of the Northern Boreal Forest. In association with red spruce are found pin cherry, aspen species, northern ash, balsam fir, and red pine. The Northern Hardwood displays sugar maple, yellow birch, and beech that dominate, along with red maple, wild black cherry, cucumber tree, hemlock, white pine, mountain laurel, red oak, pipevine, mountain holly, a variety of ferns, club mosses, hellebore, anemone, Dutchman's breeches, violets, and painted trillium.

In the Eastern Ridge and Valley Section, east of the mountains as well as in the hilly areas west of the mountains, on drier ridge tops and southeasterly slopes, the forest type vegetation is demonstrated by pines, hickories, black oak, scrub oak, scrub pine, black locust, redbud, mountain laurel, dogwood, pink lady's slipper, and azalea. Also as part of the Ridge and Valley Section there are steep shale banks with endemic species not found elsewhere under the xeric conditions existing here. Species such as shale onion, shale barren rockcress, shale bindweed, and shale ragwort are indicators of this unique plant habitat.

Within the Central Hardwood Forests are Cove Hardwoods. On north-facing slopes and in more moist habitat are found tulip tree, black and sugar maples, beech, sour gum, hickory, slippery elm, sourwood and striped maple. Shrubs such as spicebush, buffalo–nut, white alder, pawpaw, and crossvine in the southwest counties are prevalent. Spring wildflowers are abundant in this damp, hilly section of coves and north–facing slope. These include Solomon's seal, trillium, dwarf larkspur, hepatica, wild ginger, bloodroot, blue phlox, foam flower and many other species.

Within the types of forest vegetation discussed above, there are plants that reflect and are indicative of the habitat in which they are found. Swamp Forest are areas of poor drainage where the substrate is made of of a spongy organic peat–like mass formed from decayed plant material. Such an area is Cranesville Swamp in Preston County and extending into Maryland. Here in this part of the Allegheny Mountain and Upland Section of West Virginia is the only known native colony of larch or tamarack in the state. Canada yew, red spruce, yellow and sweet birch, pin cherry along with cranberries, golden club, sundew and orchid species are found. Stands of pitcher plant may be seen here but are not native species.

Heath Barrens are mountain–top areas (such as Dolly Sods) covered with a variety of heath shrubs and other low–growing plants. Chokeberry, blueberry, cranberry, huckleberry, speckled alder, azaleas, goldthread, three–toothed cinquefoil, bleeding heart, dwarf cornel, fireweed, pearly everlasting, oceannorus, wood lily, bog clubmoss and interesting grasses and sedges are a part of the Allegheny Mountain and Upland Section.

Sphagnum Glades or "bogs" are poorly drained treeless areas where sphagnum moss is dominant in association with heaths, cotton grass, bog–rosemary, sundew, orchid species, bog goldenrod, cranberry, grasses and sedges. In the Allegheny Mountain and Upland Section, Cranberry Glades in Pocahontas County is well–known.

Prairie Situations, in which are found big bluestem and little bluestem and Indian grass, are not restricted to any given areas. Artificial prairies have resulted from cultivation, allowing species from the west to move in. Among such plants are orchard grass, red clover, rye grass, ironweed, milkweed, black–eyed susans, prickly pear, cup plant and many other species.

Shale Barrens occur in the Eastern Ridge and Valley Section, particularly in Greenbrier, Hardy, Monroe, and Mineral counties. Found on hard Shale rocks of the Devonian Age and crumbling rock fragments are plants endemic to such shale conditions. Of interest are leatherflower, Kate's Mountain clover, shale onion, shale bindweed, shale ragwort and shale barren rockcress.

There are no natural lakes in West Virginia but many farm ponds, slow flowing streams, low wet areas and beaver dams are host to Aquatic Plants. Submerged species such as pond weed, milfoil, coontail, water–weed, lotus, water lily, and arrowhead are found. Around the perimeter of these ponds and wet stream edges may be seen cat–tail, burrweed, bulrush, water plantain, and sometimes in the water, water willow.

Located at or near the area where spruce forests border on the Northern Hardwoods are treeless mountain tops known as Grassy Balds, covered by oat–grass, hairgrass, redtop, and later invaders such as summer bluets, lion's paw, sheep sorrel, bracken fern, hay–scented fern and chestnut oak. In West Virginia balds occur between 3,800 and 4,800 feet in altitude. The reason for their occurrence is not known.

Throughout the state along rocky river banks, such as along the Gauley, Tygarts Valley, Greenbrier, and Cheat rivers, vegetation exhibits plants that are found only along Rocky River Bank conditions. Species such as riverbank goldenrod, Barbara's buttons, cordgrass, bladdernut, stiff aster, buttonbush, and glandular false dragonhood may be noted.

Flood Plain communities throughout display stream–side trees such as black willow, sycamore, box elder, honey locust, and more southerly, cottonweed, swamp white oak, green ash and river birch.

Ferns and fern allies (club mosses) are found throughout the state in habitats that satisfy their need for growth. In dry to moist woods, open fields, rock cliffs, moist thickets, caves, swamps, and shale, ferns find their places.

In conclusion, it is hoped that West Virginia native plant species will not be overcome by exotic and foreign invaders that are making their presence known in various areas of the state. The survival of plants and bird life as they now exist at the end of the twentieth century will depend on man's protection and diligence.

References

Core, Earl L. 1966. *Vegetation of West Virginia.* McClain Printing Company, Parsons, West Virginia.

Core, Earl L. 1975. *The Wondrous Year: West Virginia Through the Seasons.* Seneca Books, Inc., Grantsville, West Virginia.

Strausbaugh, P. D. and Earl L. Core. 1977. *Flora of West Virginia.* Seneca Books, Grantsville, West Virginia. (Now available from West Virginia University Press, Morgantown.)

History of the Brooks Bird Club

Nevada Laitsch

The Brooks Bird Club was organized at Oglebay Park in Wheeling, West Virginia in 1932 as a result of participation in a nature program provided by Oglebay Institute. A.B. Brooks, the park naturalist led Sunday morning nature walks designed to teach the public more about natural history. These walks often attracted more than a hundred people. There were special interest groups for botany, astronomy and ornithology. There was even a nature poetry group. The bird study group is the lone survivor.

By September, 1932 the desire to learn more about birds and bird life became so strong that a small group, including twelve men and one teenage girl, decided to go on their own and form a club to concentrate on bird study and go beyond the confines of Oglebay Park. The Brooks Bird Club was named for A.B. Brooks, their mentor and first naturalist of Oglebay Park.

The first meetings were held in homes of the members. However the club soon prevailed on Mrs. West, the mother of one of the members, for use of a basement room. It became known as the 'Duck Inn' due to the height of the entrance. Here regular meetings and discussions took place. Within the year *The Redstart* , the club's first publication, made its debut. It consisted of a single page which was compiled, edited and mimeographed by one man. It has grown in size and scope to become the oldest and most respected ornithological publication in West Virginia.

The club soon outgrew the 'Duck Inn' and after a time moved to 707 Warwood Avenue which has been the home of the headquarters chapter since 1944.

The club went statewide in 1940 to hold its first week–long study, called the Foray, at Lost River State Park in Hardy County. This yearly event continues to take place each June at various locations in West Virginia. The purpose is to study and document all phases of natural history throughout the state and a full report of each Foray is published in *The Redstart*..

In 1942 the club launched a second publication, a newsletter known as *The Mailbag*. This linked the club with its members serving in the armed forces and kept them informed on happenings back home. *The Mailbag* became so popular and served its purpose so well that it has been continued ever since. This now quarterly publication is sent to all members, keeping them abreast of club activities and news.

Club members began to recognize a need to explore places where little or no study had been done and where no camps were available to accommodate the numbers of people attending Forays. A decision was made to try the studies as a "camp–out" venture with each person being responsible for his or her own well– being. Over the Memorial Day weekend in 1966 fourteen persons initiated the event at Kanawha State Forest near Charleston. The Sortie, patterned after the Foray, is now an on going event scheduled for June of each year. It has evolved from a "camp–out" to the use of accommodations in West Virginia state forests or other available location. Due to facilities and availability of food, the Sortie is limited to fewer than 20 participants. Results of these studies are published in *The Redstart*.

The Brooks Bird Club has a membership of approximately 500, is non–profit, and all the work is performed on a volunteer basis.

The club has three chapters: the headquarter chapter in Wheeling, the Handlan chapter in Charleston, and an affiliate chapter, the Bibbee Bird Club, in Princeton. Each chapter implements its own programs.

The headquarters chapter runs the business end of the club. Here, regular Tuesday night meetings are held, records are stored, correspondence attended to, mailings (including all publications) are made and tasks too numerous to mention are performed. This is all done by a few dedicated members who volunteer for the good of the organization.

Overall activities of the club include: Singing Male Census since 1948, Breeding Bird Survey since 1964, North American Migration Counts, Audubon Christmas Counts, Bird Banding on Allegheny Front Mountain, Annual Hawk Counts and Spring and Fall Waterfowl Migration Counts. Yearly trips are made to Crane Creek and Magee Marsh in Ottawa County, Ohio and to the Eastern Shore in November. Special projects include a scholarship program and participation in the Breeding Bird Atlas.

Calendar of Events
The Brooks Bird Club

Date	Activity	Place
January	BBC Membership Month	Wheeling, WV
February	Mid–Winter Meeting	Weston, WV
March	Waterfowl Trip	SenecaLake,WV
April	Old Hemlock weekend	Preston County
May	Ottawa Warbler Trip	OakHarbor,OH
June	Breeding Bird Survey	West Virginia
June	Annual Foray	West Virginia
June	Sortie	West Virginia
Aug 15 Oct 1	Bird Banding –Dolly Sods	Red Creek, WV
September	Labor Day Weekend	West Virginia
September	Hawk Counts	WV Mountains
October	Cedar Lakes Annual Mtg	Ripley, WV
November	Board of Directors Mtg	Wheeling, WV
November	Eastern Shore Trip	Eastern Shore
December	Christmas Bird Count	West Virginia

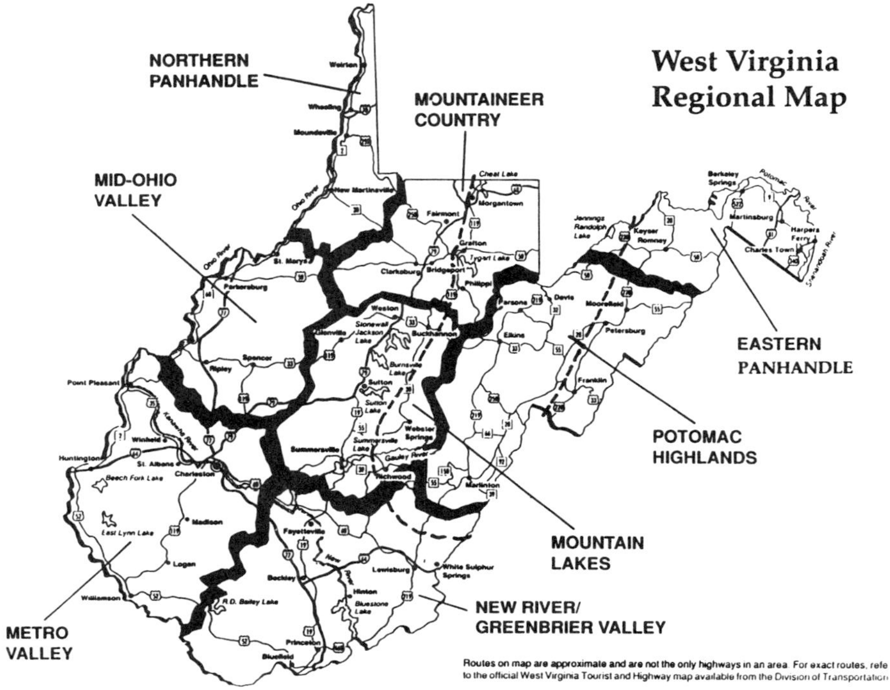

Fig. 2. Map of hot spot regions.

Part Two

Birding and Botanizing Hot Spots

Birding Hot Spots are organized in sections by region that share similar habitat and topography. The map on the facing page defines the area covered by each region.

The solid lines divide the birding hot spots into geographical regions:

1. **Eastern Panhandle** bordered by the Potomac and Shenandoah rivers.
2. **Potomac Highlands** containing highest elevations and wilderness areas.
3. **Mountain Lakes** in the central part of the state.
4. **New River and Greenbrier Valley** comprising southeastern counties.
5. **Metro Valley** including the population centers of Charleston and Huntington.
6. **Mid-Ohio Valley** centered on Parkersburg and the Little Kanawha River.
7. **Northern Panhandle** containing the Wheeling and Weirton areas.
8. **Mountaineer Country** centered on Morgantown and the Monongahela River.

The articles on individual birding hot spots in this section are grouped in the appropriate geographical region.

Birds of Particular Interest

Wendell Argabrite

Golden Eagle – The Golden Eagle can usually be found around the stripmined areas of Cheat Mountain in the winter.

Virginia Rail – The Virginia Rail can easily be found at Greenbottom Wildlife Management Area in the spring and early summer.

Common Moorhen – In the spring and summer, the Common Moorhen can usually be seen at Greenbottom Wildlife Management Area.

Northern Saw–whet Owl – One can usually hear the Northern Saw–whet Owl on top of Cheat Mountain in early spring.

Prothonotary Warbler – Greenbottom Wildlife Management Area offers the best chance to find the the Prothonotary Warbler.

Swainson's Warbler – The Swainson's Warbler can usually be found in the southern counties, such as Logan, Mingo, etc. Also see the specific writeup for the Swainson's Warbler.

Northern Waterthrush – Cranberry Glades Boardwalk offer easy listening and viewing of the Northern Waterthrush.

Mourning Warbler – The Mourning Warbler can usually be found in the high mountains, Gauley Mountain being one of thee best places.

Canada Warbler – Cranberry Glades is one of the best mountain areas to find the Canada Warbler but also along the Arboretum Trail at Watoga State Park.

Blue Grosbeak – The Blue Grosbeak is usually found at the Robert C. Byrd Locks and Dam area and the Greenbottom Wildlife Management Area.

1. Eastern Panhandle

Bob Dean

West Virginia's three eastern-most counties provide a smooth transition from the Piedmont of Virginia to the ridge and valley of the Appalachians. As one crosses the Blue Ridge Mountain into Jefferson County the broad expanse of the **Shenandoah Valley** spreads out to the west. Here at historical Harpers Ferry the Potomac and Shenandoah Rivers converge to make passage through the Blue Ridge and into Maryland and Virginia at one of the most famous water gaps in America. (Birders attempting to enhance a state list of birds should be aware that the Potomac River belongs to Maryland and plan accordingly.)

The Shenandoah Valley is home to Charles Town and Martinsburg, and its proximity to the Washington-Baltimore area is producing tremendous population growth at this time. The impacts on all of the sites detailed here and the region in general are likely to be profound in the next several decades unless measures are taken to identify and maintain significant natural areas and offer reasonable protection from the effects of overuse and incompatible land use.

In southeastern Jefferson County is **Shannondale Springs Wild Life Management Area**. Just west of Charles Town is the unique **Altona Piedmont Marsh**. The next mountain to the west is **North Mountain** which extends in broken segments far to the north and south. The town of Hedgesville lies in a major gap of North Mountain where SR 9 crosses. To the west of North Mountain is **Back Creek Valley** which is a good birding area in all seasons. From SR 9, there is one main artery through the valley, CR 7 to the south. It passes through the communities of Tomahawk, Jones Spring, Shanghai, and Glengary as it meanders through an abundance of good birding areas. **The Sleepy Creek Wildlife Management Area** is on the west side of Back Creek Valley and stretches nearly from the Virginia line to SR 9.

The historic Town of Bath (Berkeley Springs) is in the next valley to the west and is deserving of at least a lunch or dinner stop while in the area. **Cacapon State Park** is about 9 miles south of Berkeley Springs on the west side of US 522. It is a wonderful spot for staying the night or a full vacation, and has numerous recreational activities from swimming in the lake to horse back riding or hiking on the Cacapon Mountain trails of the park.

Good birding can be found at all of the areas mentioned here, but the "bird of the day" might as easily be located while traveling from one hot spot to another. They are where you find them, and the Eastern Panhandle is a fine place to look.

Altona Piedmont Marsh

Bob Dean

The nearly 100 acres of the **Altona Piedmont Marsh** serve as the ultimate example of a marl wetland in West Virginia. Through the extensive efforts of The Nature Conservancy and the extreme generosity of the owners, this unique area has been preserved for future generations.

Access to the marsh is from SR 51 about 1.7 miles west of Charles Town. Turn north from SR 51 at the historical marker for "Ruins of St. George's Chapel." Park either at the ruins on the left or just before the railroad track on the right. This is private land so be respectful. Proceed to the right on the railroad track. The RR is active. Good birding often begins at that point. Evitts Run, the main water supply for the area is on the left, but soon crosses under the track to the right where it meanders through the marsh until departing at the east end into agricultural land. As visitors walk down the track they will pass through wooded swamp, cattail marsh, alder-willow swamp, sedge-rush marsh, and then an open water area and more cattails and woods before the marsh ends.

The vast majority of marsh birds found in the state have been recorded here at some time. Many are still commonly found in the proper season. In winter one can often find Common Snipe, Northern Harrier, Red–headed Woodpecker, Fish Crow, Swamp Sparrow, and Rusty Blackbird. Late May to mid–June is often the most rewarding time. Migrants and breeding birds are both arriving in numbers. American and Least bitterns, Green Heron, Virginia Rail, Solitary and Spotted sandpipers, Marsh Wren, Loggerhead

Significant Birds of Altona Piedmont Marsh

American Bittern
Least Bittern
Great Blue Heron
Green Heron
Black–crowned Night Heron*
Northern Harrier
Northern Bobwhite
King Rail*
Virginia Rail
Sora
Common Moorhen*
Common Snipe
American Woodcock
Short–eared Owl*
Red–headed Woodpecker
Willow Flycatcher
Fish Crow
Sedge Wren*
Marsh Wren
Loggerhead Shrike
Rusty Blackbird
Blue Grosbeak
Bobolink*

* seen prior to 1980, but not on recent list

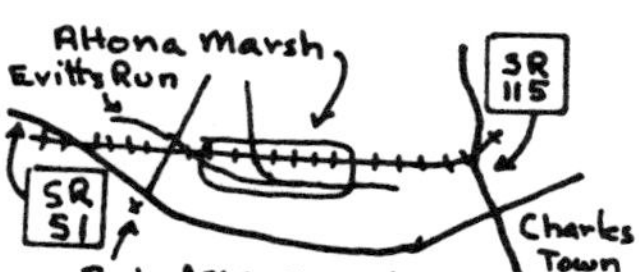

Significant Plants of Altona Piedmont Marsh

Water Pennywort
Hard-stem Bulrush
Water Smartweed
Water–horsetail
Common Skullcap
Baltic Rush
Suberect Sedge
Bearded Sedge
Wooly Sedge
Whorled Loosestrife
Cone–shaped Sedge
Lake Sedge
Tufted Loosestrife
Winged Loosestrife
Buxbaum Sedge
Hairy–fruited Sedge*
Prairie Sedge*
Rigid Sedge*
Intermediate Sedge
Kalms Lobelia*
Swamp Lousewort
Large–leaved Grass–of–Parnassus
Small–beak Sedge*
Flat–top Joe–pye Weed*
Thin–leaved Hedge Nettle
Sterile Sedge*
Holy Grass
Tall Spikerush

* found only at Altona Piedmont Marsh in WV

Shrike, Warbling Vireo, and a good variety of warblers can all be found now.

Birds are far from being the only significant inhabitants of this wonderful marsh. The spotted turtle and the northern cricket frog are two uncommon herps found at Altona. There are also a number of rare plants found in the marsh. At least seven are found nowhere else in the state.

The accompanying list of both bird and plant species to be seen at Altona Piedmont Marsh (see box) should be incentive enough to set the trip planning gears in motion. It is suggested that people wanting to visit Altona Piedmont Marsh do so in small groups on caravan from Charles Town as parking is somewhat limited. Please stay on the railroad

Sleepy Creek Wildlife Management Area

Bob Dean

Sleepy Creek Wildlife Management Area is located on the Berkeley–Morgan County line in the eastern panhandle of West Virginia. Its nearly 23,000 acres afford it the status of being the largest contiguous parcel of state–owned land. It was purchased by the state in 1951 as a state forest, and has been a wildlife management area since 1954. Over two- thirds of the acreage is in Berkeley County. Only the western slope of Sleepy Creek Mountain and the northern tip lie in Morgan County.

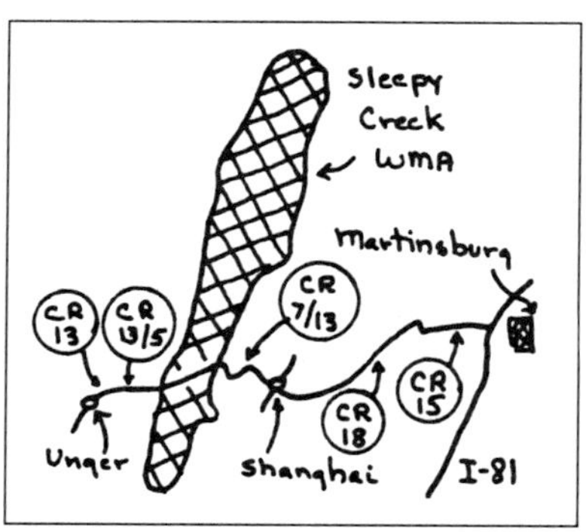

To reach Sleepy Creek Wildlife Management Area from the east, at the I–81 and CR 15 exit near Martinsburg, proceed west on CR 15 (which becomes CR 18) for 10 miles to Shanghai.

Then take CR 7/13 to the Sleepy Creek WMA. From Morgan County, one must take CR 13 and CR 13/5 from Unger.

Sleepy Creek lies on a north-south axis and is approximately 17 miles long and one to three miles wide. It contains most of two mountains: Sleepy Creek to the west, the top of which is the county line; and Third Hill to the east, which traverses its entire length. The two mountains join at what is locally referred to as **"The Locks"** about five miles from the southern end of the area. The highest point in the WMA is High Knob (2,172 feet above sea level), which is on Third Hill Mountain just south of the only access road to cross the area from east to west. This road, known locally as Shanghai Grade Road crosses at "The Locks. " The lowest elevation (about 630 feet) is near the small community of Shanghai on the east side.

Sleepy Creek is located in the Ridge and Valley physiographic province, with its characteristic trellis drainage pattern. The valley between the two mountains and south of "The Locks" is in the Brush Creek watershed, which drains into Back Creek to the east after leaving the WMA. The birder's gem of the area is the 205–acre **Sleepy Creek Lake** and adjacent wetlands. The dam is located about six miles north of "The Locks" on Meadow Branch and the lake is about 2.5 miles long. Camp sites are available.

There are about 200 acres of clearcuts and wildlife food plots on the area. With the exception of a small hemlock-white pine forest in the north end, the rest is mixed hardwood and oak–pine forest. Much of this is undergoing some degree of transition from the recent heavy oak kill by gypsy moths and a combination of other factors.

Spring Migration — Early May is prime time for birding here just as it is throughout the Middle Appalachians. At this time birders can readily find 80–85 species here. On a day of heavy migration the area around the lake and the mountain tops will be teeming with mixed flocks of warblers and vireos feeding wherever the sun is bright and the oaks are blooming. It's not uncommon to list five species of vireos and over twenty species of warblers. There are few West Virginia warblers that cannot be found here during migration, the Prothonotary and Swainson's being the only absentees. The Warbling Vireo is the only local vireo not found here. In late spring it is not uncommon to find a few lingering Ruby–crowned Kinglets.

The spotted thrushes migrate through the area in fair to good numbers from year to year. The Gray–cheeked is rare while the Hermit, Swainson's and Veery can often be found on the west side of the lake or near "The Locks." These are also two of the best places to find some of the last Dark–eyed Juncos and White–throated Sparrows to depart for their summer territories to the north. The Wood Thrush is the only common spotted thrush at this season. The American Robin and Eastern Bluebird are both fairly common in the lake area as well as where the WMA borders appropriate habitat.

In early spring there are generally some waterfowl present at the lake after thaw. Many species of ducks can be found here from time to time, but the most common are Wood Duck, Mallard, Pintail, Lesser Scaup, Blue–winged Teal, and the three mergansers. Other water birds that frequent the lake during this season include Common Loon, Pied–billed and Horned grebes, Double–crested Cormorant, Canada Goose, Ring–billed and Bonaparte's gulls, and American Coot. Spotted and Solitary sandpipers often stop in migration. Green and Great Blue herons are regular visitors once they've reached the area.

Summer Residents — A typical summer dawn will greet the Sleepy Creek birder with a wonderful chorus highlighted by Whip–poor–wills, Barred Owls, Wild Turkeys, Wood Thrushes, and Scarlet Tanagers. Depending on exact location they will be accompanied by some of the sixteen warblers species (See page 136), the three vireos, five woodpeckers, five flycatchers, two chickadees, three blackbirds, and three sparrows that nest in the area. There is a significant concentration of cavity nesters at the upper (southern), wooded end of the lake, including numerous Tree Swallows, Eastern Bluebirds, Great–crested Flycatchers, Black–capped Chickadees, Tufted Titmice, White–breasted Nuthatches, Wood Ducks, and several woodpeckers. There will also be Baltimore Orioles, Eastern Kingbirds, Blue–gray Gnatcatchers, and many others nearby. At least eighty-two species of birds have been confirmed as breeders on the WMA.

Many acres are being managed as Ruffed Grouse habitat. This, along with the areas of extensive gypsy moth kill, has created a boon for this species and their drumming is now a very common and welcome part of the birder's day.

Great Horned, Barred, and Eastern Screech owls are all permanent residents here and can be heard at predawn or dusk in many areas. There are several species of hawks that nest here including Sharp–shinned, Cooper's, Broad–winged, Red–shouldered, and Red–tailed. There are two Osprey nests at the lake. The first was started in 1986 and has been used since 1987. The second nest was built in 1992 and has been used since 1993, and is about one mile down lake from the first. One of the adults from the first pair was a bird that was hacked at the South Branch of the Potomac site by the Non–game Wildlife and Natural Heritage Program of the West Virginia Department of Natural Resources.

Both vulture species have nested here and can be observed from the top of **Third Hill Mountain**, entering nest sites in the valley below the lake. All three large WV corvids can be observed, the Fish Crow being most common in the north end and along the eastern edge of the WMA.

Fall Migration — The fall season is a mixed bag. At times there will be numerous mixed flocks of migrants seemingly everywhere. At other times it's difficult to find any birds other than resident woodpeckers, titmice, corvids, etc. The best time is generally after a cold front has gone through in mid to late September.

By late in the season some water birds have generally begun to appear on the lake. Most species that are present in the spring will also be present to some degree in the fall. It's a hit or miss proposition, and one that is well worth the effort when you catch it right.

There is a **fire tower** on top of **Sleepy Creek Mountain** that used to extend well above the tree tops, but in 1993 it was torn down to a level below the trees. Local birders had some very good hawk watches from this tower before it was lowered. The tallies resembled those of other Appalachian ridges, with Broad–wing Hawks leading the list followed by Red–tails, Sharp–shinned hawks, and enough assorted Ospreys, Red–shoulders, Cooper's, Merlins, etc. , to keep one's eyes skyward. The hawks are still there, but it requires a little more effort to find the right rock outcrop or other vantage point to benefit from the current conditions. Most of the hawks are funneled in from mountains to the north of

Maryland and Pennsylvania that have been gapped by the Potomac River.

Winter — If winter is a season of reflection then Sleepy Creek WMA is a great place to reflect. Birding is generally slow at this time of year, but that helps to justify more time in the car and close to the heater. Birding tends to take on more than the usual social tilt.

At times one can drive for miles and not see a bird, but there are always centers of activity. It's our job as industrious birders to locate them. Perhaps a secluded pine thicket on the east side of Third Hill mountain is teeming with kinglets, chickadees, nuthatches, creepers, and an assortment of woodpeckers. Perhaps the season has been mild and the lake is not yet frozen. There will likely be an abundance of late season water birds, such as Common Loons, Pied–billed and Horned grebes, and a wonderful mixture of ducks. Perhaps the local Barred Owl chorus will do a matinee performance only to be joined by a pair of Great Horned Owls. Perhaps there will be a late season hawk movement with several eagles for spice. It is a very unpredictable season, but one that can yield its share of unexpected highlights. Do yourself a favor and visit Sleepy Creek Wildlife Management Area whenever the opportunity arises. Blue–winged and Golden–winged warblers are suspected of breeding, are seen during breeding season but not yet confirmed.

Maps are available by to West Virginia Wildlife Resources Division, 1 Depot St. Romney, WV, 26757.

Shannondale Springs Wildlife Management Area and the Shenandoah River

Bob Dean

Nestled between the Shenandoah River and the Blue Ridge Mountain on the eastern edge of West Virginia is the 1350 acre **Shannondale Springs Wildlife Management Area**. The area is in two separate parcels. The lower 610 acres is comprised of a large neck of the Shenandoah and a high ridge that extends to the south. The upper 738 acres is mostly recently select cut mixed hardwood in an interesting assembly of small drainages between the ridge above the river and the Blue Ridge Mountain. The eastern boundary extends to about the 1300 foot elevation.

To get to Shannondale Springs WMA take SR 9 approximately 1.2 miles east of the Shenandoah River to CR 9/5 or Mission Road. Turn south 2.0 miles to the first of two parking areas on the right for the lower section. The first road follows the river on the north side of the WMA. The second hops over the high ridge in the center of the lower section and then provides several options along the way before converging with the first road.

Approximately 1.7 miles further south is a parking area on the east side of Mission Road which accesses the upper section of the WMA. Maps may be obtained by writing: West Virginia Wildlife Resources Division, 1 Depot St. Romney, WV, 26757.

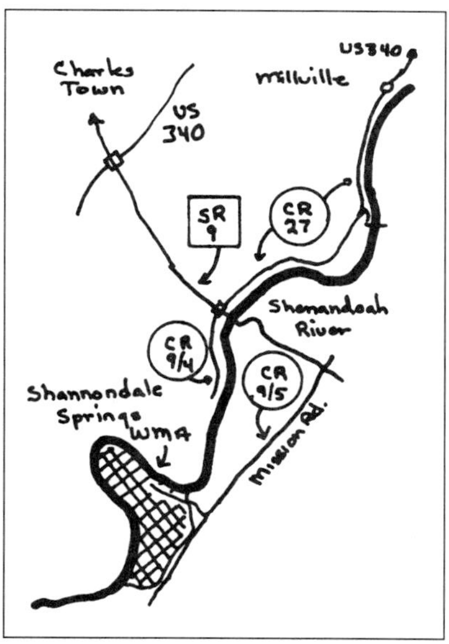

The entire area can be covered fairly thoroughly in a day of birding. Some of the trails have brief steep sections, but overall the slope of the ridges is moderate to level.The trail network is very well developed and maintained in the lower section, but not very extensive in the recently acquired upper section.

The spring migration brings many species to Shannondale Springs: Horned and Pied–billed grebes, Green Heron, Great Blue Heron (year–round), Black Vulture (year–round), Spotted and Solitary sandpipers, Belted Kingfisher (year–round), all six species of swallows, Fish Crow (year–round), Solitary Vireo, and most expected warbler species. Some of the migrants that will stay for the breeding season are the Yellow–billed Cuckoo, Whip–poor–will, Wood

Thrush, White–eyed, Yellow–throated, Red–eyed and Warbling vireos, Blue–winged, Prairie, Cerulean, and Worm–eating warblers, Northern Parula, American Redstart, Yellow–breasted Chat, Scarlet Tanager and Baltimore Oriole just to name a few.

The fall migration can be rather spectacular at times also, mixed flocks of perching birds keeping binoculars busy. But the most interesting time for birding here is late fall and winter. If the river is not frozen, there is often a good variety of waterfowl, including Bufflehead, American Wigeon, both teal species, Wood Duck, Common Goldeneye, and all three mergansers. In brushy areas along the river, particularly on the south side of the WMA, one can usually find Winter Wren. Above the river one can usually find most of the expected over-wintering species including Yellow–bellied Sapsucker, both kinglets, Hermit Thrush, Myrtle Warbler, Eastern Towhee, American Tree, Field, Song, White–throated, White–crowned and occasionally Swamp and Lincoln's sparrows. Some winters will bring Purple Finches and Evening Grosbeaks although this in increasingly rare. They are more often seen in early spring or late fall, often just flying over.

Elsewhere along the **Shenandoah River** birding events can be even more dramatic. The spring flights of warblers from the Virginia line to Harpers Ferry can still be downright impressive. Although Yellow–throated and Prothonotary warblers are rarely encountered at Shannondale Springs they are annual breeders downstream a few miles. In early May all six swallow species have been observed in the chilly early morning hours on the power lines that cross the river just a few hundred yards below the SR 9 bridge. But for sheer numbers of swallows one must see the fall migration to appreciate breeding success and congregation at a food supply. At times the air will be so full of birds for some height above the water that collisions would seem almost inevitable.

In addition to the waterfowl found regularly at Shannondale Springs there are others seen between there and Harpers Ferry, particularly in winter. Add the Northern Shoveler, Gadwall, Canvasback, Redhead, both scaup species, and the Ruddy Duck to the list. If there is an unusual invasion of gulls or other water birds count on them to show up here. For example, there were several Red–necked Grebes here during the late winter and early spring of 1994.

There are approximately 17 miles of the Shenandoah River in West Virginia. Take SR 9 east of Charles Town and turn either south or north just before the bridge. To the south is CR 9/4 which dead-ends approximately 0.8 miles upstream. To the north is CR 27 which parallels the river to Millville then diverges to US 34 just west of Harpers Ferry. Approximately 2.5 miles north of SR 9 is a steep paved access road over a small ridge to the **Millville dam**. This is a must in spring, fall, or winter if the river is open.

2. Potomac Highlands

Greg Eddy

The Potomac Highlands include the highest elevations in West Virginia. This area also covers many of the most popular tourist destinations in the central Appalachians. These include; **Canaan Valley, Dolly Sods, Sinks of Gandy, Spruce Knob, Gaudineer Knob,** and **Seneca Rocks**. The high elevations attract many species seen only in this part of the state and many that breed only in this part of the state. These include Alder Flycatcher, Dark–eyed Junco, Veery, Swainson's and Hermit thrushes, Blackburnian, Canada, Magnolia and Mourning warblers, Northern Waterthrush and even Nashville and Yellow–rumped warblers.

Because of the beauty of the area and the species present, this is the most popular birding area in the state. Most of the prime areas are covered in the articles that follow. But some additional interesting places can also be visited when visiting the Potomac Highlands, such as the **National Radio Astronomy Observatory** and the **Cass Scenic Railroad**, as well as many caves.

Alder Flycatchers are found in several places in the Highlands and in the greatest concentration in **Canaan Valley**. But another interesting location is on **Middle Mountain** at the intersection of FSR 14 and Little River Trail. Although Cliff Swallows are not common in WV they are found in several parts of the state. An interesting site is a barn in the middle of a field north of and within sight of Bartow on US 250 in Pocahontas County. Another location is in the agricultural sale barn on the north side of Terra Alta in Preston County. A fairly reliable location for Loggerhead Shrike is on **Back Mountain Road** (CR 1), which intersects US 219, east of Gaudineer. They are usually found in a sheep pasture on the west side of the road three or more miles from Durbin.

The headwaters of **Shavers Fork of the Cheat River** is an area that continues to surprise birders. The Southernmost breeding record for White–throated Sparrow comes from there and Goshawks and Olive–sided Flycatchers have also been found there in the breeding season. The area is south of

Cheat Bridge and Barton Knob discussed in the Gaudineer/ Barton Knob article.

Other fine places to bird and botanize are in Fernow Experimental Forest and Olson's Bog on Backbone Mountain, both outside of Parsons and Kumbrabow State Forest.

Lost River State Park

Jane Whitaker

Lost River State Park is located in Hardy County, on the dry eastern slopes of the Alleghenies. From Baker on SR 55, proceed south on SR 259 to Mathias. Turn west on CR 12 and go 4 miles to the park entrance.

There are not only a wide variety of birds at the park but also an abundant flora and fauna including all three species of native squirrels, skunk, raccoon, white–tailed deer, bobcat and black bear. Herpetologists will find a variety of amphibians and snakes as well. In early to mid-June the mountain laurel and rhododendron throughout the park will be in bloom.

Start your tour of the park at the Superintendent's office where you will be able to pick up a bird list and a trail map. In winter, be sure to check the feeder near the building for Downy Woodpeckers and Black-capped Chickadees.

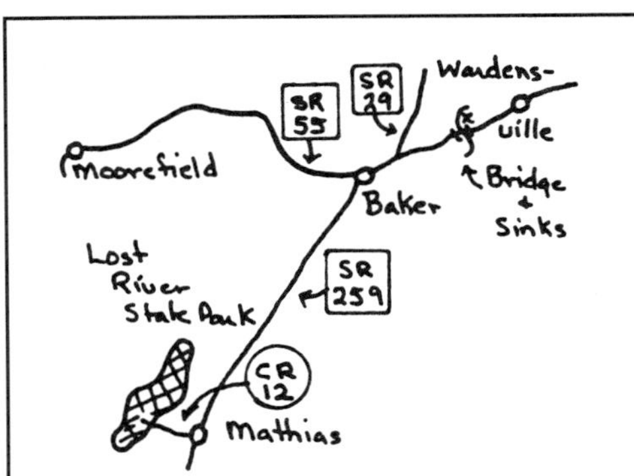

As you leave the park office check the creek near the main road. You should find Acadian and Least flycatchers, and possibly a Louisiana Waterthrush. Eastern Wood–Peewees, vireos and Great Crested Flycatchers can be heard throughout the park. Turn left from the parking lot and drive through the cabin area. About 30 yards up the road on your left, there is cancer root and possibly Indian pipe in the woods. If you drive to the turnaround, 0.8 mile, and park, the **Loblolly Trail** has a very gentle slope for several hundred yards which is easy walking. Blueberries, huckleberries, trailing arbutus and a variety of ferns, including some very rare ones, can be easily found. Listen for Hooded Warbler and other warblers as you go.

Retrace your steps to the Park office and continue to the swimming pool parking lot. American Redstarts and Northern Parulas can sometimes be found here and it's not uncommon to hear the whistle of a Broad–winged Hawk overhead. From the parking lot you can look over the pool and see Cranny Crow Lookout on Big Ridge. Ravens nest there and can be seen easily all year round.

Cross the road in front of the parking lot and look for Eastern Phoebes nesting under the bridge. The trail leading to the **Lee Sulphur Springs** winds through a small stand of virgin forest with giant hemlocks and rhododendron. The water is still used by the local people and if you have never experienced sulphur water you should take a sip. The smell is worse than the taste.

Return to your car and continue past the swimming pool, making the first left turn. There is a tree in the middle of the road, so it is hard to miss. Continue 0.25 miles to the recreation field. At the upper end of the field there are a number of 2 to 3 foot high Allegheny ant mounds. They can also be seen at the park boundary line 0.2 miles further down the road. Return to the main park road and proceed toward Mathias. Be sure to note the fire tower road (gated at the bridge) which leads to Big Ridge and Cranny Crow Lookout. It is 0.05 miles past the Dove Hollow Road that leads to the stables.

Side trip to **Big Ridge** — An interesting hike to Big Ridge will take about 1/2 day (3 to 4 mile round trip) of walking. There are two trails which lead up the mountain as well as the fire tower road which is shorter. The Big Ridge Trail is preferable despite its longer length because it is neither as steep nor as frequented by horseback riders. In late winter Ravens are easily seen from Cranny Crow Lookout. In Spring (early May), you will be serenaded as you walk by flocks of Rose–breasted Grosbeaks, Black and White Warblers, Eastern Towhees, and Great Crested Flycatchers. Ruffed Grouse are often flushed near the top of the mountain, and Wild Turkeys are in the area, but hard to see. In May, gaywings and vernal iris can be seen along the trail. The top of the mountain was farmed as early as 1840. It is still mowed by the park, so an interesting contrast in flora and fauna can be seen there. Bald Eagles and hawks have been seen from the lookout. The Allegheny ant mounds are present here too. Scrub oak, chinquapin, and table mountain pine are plentiful along the upper sections of the trail.

Shale Barrens — Upon leaving the park be sure to observe the extensive shale barrens on your left just before the intersection with CR 259 at Mathias. They run for at least 100 yards. Cactus and other shale barren plants such as shale barren pussytoes can be seen from the road. Turn left onto CR 259 and notice the fossil ripple marks in the shales in the first outcrop on your right.

Lost River Sinks and Hawk Recreation Area

Jane and Matthew Whitaker

Rough-winged Swallows can be seen flying in out of the rock crevices where they nest within a few miles of the park. To make a side trip (see Lost River State Park map) for these birds follow SR 259 north to Baker (13.5 miles). Turn right onto SR 55 and proceed about 2 miles and stop at the abandoned quarry on your left .

Proceed another 5.2 miles and stop at the road side park beyond the bridge. **Lost River Sinks** are to your left. The sink can be seen from the path leading to the river across the road from the park. Lost River flows underground and emerges on the other side of the mountain near Wardensville where it acquires a new name — the Cacapon River.

Be sure to listen for both Least and Acadian flycatchers at this stop. Proceed on SR 55/SR 259 for 4.5 miles to Wardensville and bear right on SR 55 where it splits from SR

259 at the far side of town. Travel 4.3 miles, going up a steep mountain. Look for an abandoned roadside park on your right. The secondary road across from the abandoned park leads to **Hawk Recreation Area**. Turn left and proceed slowly for 4 miles, birding along the way. There are Ovenbirds, Scarlet Tanagers, Eastern Towhees, Cedar Waxwings, vireos, warblers, woodpeckers, Common Ravens, and several species of raptors throughout the forest. Great Crested Flycatchers are common. Clear-cut areas will have Blue–gray Gnatcatchers, Prairie Warblers, Indigo Buntings and Field Sparrows. Ruffed Grouse and Wild Turkeys are often seen along the road.

Turn left at the first road (about 4 miles). It will probably be without a sign but it leads to **Hawk Campground and Recreation Area** — about 1 mile. There is a picnic area with pumps for water and pit toilets. Many small logging trails lead off the access road and any of them can be easily traversed. The **Big Blue Trail** intersects the picnic area and parallels the road for 150 yards. The birding during migration seems to be particularly good along this part of the trail. Sometimes Wild Turkeys can be seen here. The flora on the trail and along the road bank to the left is especially diverse.

Follow the Big Blue Trail as it crosses the road into a half grown clear cut. Continue until the trail reverses itself to the right. If you bear left here you will come to a clear cut which has Prairie Warblers, Field Sparrows, and Indigo Buntings. You may follow the logging trail here or go back to the Big Blue. Both trails will go downhill to small streams where there are flycatchers and some unusual plants, including golden-knees, violet oxalis, wild comfrey and wood betony. Give yourself about two hours for the Big Blue Trail, a much more strenuous and longer hike rewarded by a pretty waterfall where it crosses Hawk Run.

Plan to be at the campground at dusk. Prior to sunset you will be serenaded by Eastern Phoebes, White-breasted Nuthatches, Red-bellied Woodpeckers and Great Crested Flycatchers. As night approaches Wood Thrushes sing, and will often continue throughout the night. If you leave after dark you will probably see Whip-poor-wills along the secondary road that leads back to SR 55.

Canaan Valley State Park and Surrounding Area

John Northeimer

Canaan Valley State Park is located 35 miles northeast of Elkins, West Virginia and 40 miles southwest of Oakland, Maryland. From Elkins take SR 33 east to Harman, then take SR 32 north for 10 miles to the park entrance on the left. From Oakland take US 219 south to Thomas, WV, then take SR 32 for 16 miles to the park entrance on the right.

Canaan Valley State Park is located within **Canaan Valley**, the largest high valley east of the Rocky Mountains. The high elevation, diverse plant communities and varied landscape provide for extraordinary birding at a middle latitude. Six thousand acres of shrub swamp wetland, boreal forests and northern hardwood forest provide a rich mix of northern and southern plant and animal species.

Canaan Valley State Park is located in the southwestern portion of the oval–shaped valley. Fifteen trails provide access to the park interior and the valley rim. Be sure to pick up a trail map and a bird checklist at the park **Nature Center**. Guides are also available.

Spring Specialties — Due to the high elevation and less defined migration routes north, spring is a true warbler paradise within Canaan Valley. High elevation species abound. The top seven warblers in abundance during early May are Magnolia, Black–throated Blue, Black–throated Green, American Redstart, Common Yellowthroat, Hooded and Canada.

Large unbroken forest tracts produce Veery, Hermit and Wood thrush, Solitary Vireo, Scarlet Tanager and Rose–breasted Grosbeak. Agricultural lands and tall grass areas yield Indigo Bunting, Savannah Sparrow and Bobolink.

Early spring brings the *pent* call and mating flight of the American Woodcock. Woodcock are abundant in the alder thickets that line the park road adjacent to the golf course.

All of the species mentioned above are local breeders, presenting a wide window of opportunity for observation. Spring Migration Counts have recorded a cumulative total of 132 species within Tucker County.

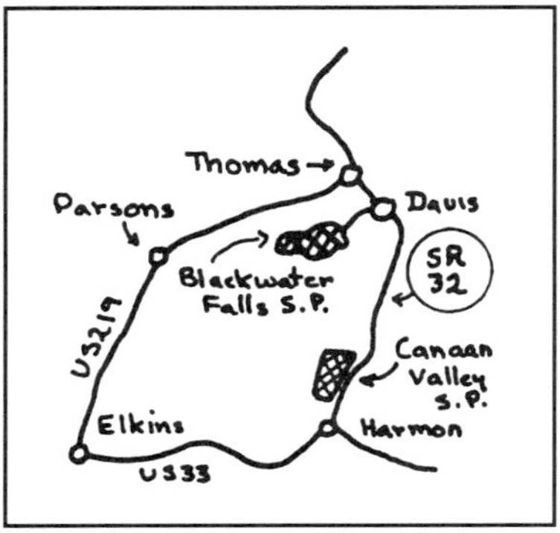

Migrating waterfowl include Common Loon, Tundra Swan, Wood Duck, American Black Duck, Blue–winged Teal, American Wigeon, Ring–necked Duck, Lesser Scaup, Bufflehead, Hooded and Common merganser.

Winter Species — High elevation and a cold climate control winter bird populations in Canaan Valley. Christmas Bird Counts have recorded a cumulative total of 61 species. What the area lacks in species diversity it makes up for in speciality. Rough–legged Hawks (both dark and light phases) are regular winter residents along with Horned Lark, Common Raven, Red–breasted Nuthatch, and Purple Finch (all permanent residents). Invasions of Evening Grosbeaks occur every 2 to 3 years. Common Redpoll and Pine Siskin often join these invasions. Common Loon, Horned Grebe and Hooded Merganser are possibilities of the Blackwater River during intervals with open water.

Abe Run Trail in Canaan Valley State Park — A walk along Abe Run Trail is a rewarding experience. This route takes you through typical shrub swamp wetlands — in fact, through four different types of wetland or wetland-associated plant communities. Speckled alder, balsam fir and eastern hemlock represent the shrub community with marsh marigold, glade spurge, marsh St. John's wort and false hellebore as ground cover. Proceeding east to west as you cross the boardwalk, you will encounter shrub swamp, wet meadow, northern bog, and a hemlock grove.
Green Heron, Pileated Woodpecker, Great Crested Flycatcher, Cedar Waxwing, Solitary Vireo, Magnolia Warbler, Common Yellowthroat, Indigo Bunting, Song Sparrow, and Swamp Sparrow are commonly encountered along Abe Run Boardwalk.

The Abe Run/Mill Run/Deer Run Trail system represents a wildlife observation area. Reserve time to explore this trail network.

Middle Ridge Trail, Canaan Valley State Park — Middle Ridge Trail is a loop trail of 2.5 miles. Primary forest cover is American beech, red and sugar maple, black cherry and yellow birch. The west slope of Middle Ridge is covered with large patches of "club moss" or Lycopodium. Painted trillium blooms in early May along this trail.

Forest management on the park precludes timber removal, including down or dead trees. Barred Owls as well as Downy, Hairy, and Pileated woodpecker take advantage of the abundant food and cavities created by this type of forest management. White–breasted Nuthatch, Wood Thrush, Red–eyed Vireo, Magnolia and Black–throated Green warbler, Ovenbird, and Scarlet Tanager are commonly heard, but seldom seen.

An alder thicket and old beaver dams at the halfway point present ideal habitat for Great Blue Heron, American Woodcock and Swamp Sparrow.

Back Hollow/Red Creek P.O./Jenningston, a Regional Driving Route — Turn right onto Back Hollow Road (CR 45) from SR 32 just south of Canaan Valley State Park entrance. Follow this for 2.5 miles to SR 72. Turn left and then right at Red Creek, continuing on CR 45 to Jenningston.

A productive driving route, particularly in the spring, this route descends from the southern end of Canaan Valley (elevation 3100 feet) to the Dry Fork River (elevation 1900 feet). The transition from mixed northern hardwood to oak–hickory forest type presents many opportunities. Bobolinks inhabit the agricultural fields on both sides of CR 45, 2.5 miles from the Red Creek Post Office.

Listen for Broad–winged Hawks as you descend through forest tracts to Jenningston. Eastern Wood Pewee, Wood Thrush, Brown Thrasher, White–eyed Vireo, Northern Parula, Chestnut–sided, Magnolia and Hooded warblers, American Redstart, Scarlet Tanager, Rose–breasted Grosbeak, Indigo Bunting, and Baltimore Oriole are in appropriate habitats along the route.

Wildflowers are abundant and in full bloom in early May, distracting even the most ardent birder with the mildest interest in botany.

Dry Fork River at Jenningston has many flat pools.

Great Blue Heron, Mallard and Belted Kingfisher frequent this section of stream. The route may be extended by following CR 35/15 to Gladwin on the south side of Dry Fork River. Follow CR 35/16 along the drainage of Big Run on the return trip. Big Run is a beautiful stream with many low cascades where you are sure to hear Acadian Flycatcher and Louisiana Waterthrush.

Blackwater Falls State Park — Turn left out of Canaan Valley State Park onto CR 32 North. Follow CR 32 for 12 miles to Davis, then turn left on Blackwater Road and travel one mile to the park entrance.

Blackwater Falls State Park has numerous trails covering a variety of habitats. Trail maps may be obtained at the Main Lodge or Nature Center.

Yellow Birch Trail, which begins at the **Gentle Trail** parking lot, winds through hemlock groves and mixed northern hardwood stands. A bog is present not more than 1/4 mile from the trail head. This acidic, sphagnum bog is representative of wetlands associated with the Pottsville Group of sedimentary rocks (mainly sandstones) which cap the ridges and high plateaus of the region.

A walk along Yellow Birch Trail will yield Black–capped Chickadee, Golden–crowned Kinglet, Hermit Thrush, Solitary and Red–eyed vireo, Black–throated Blue and Black–throated Green warblers.

Make sure that you visit the overlook behind Blackwater Lodge in the afternoon or evening to hear the downward spiral song of the Veery.

Turkey Vultures are seen all year soaring along the rim of Blackwater Canyon.

Take time to see **Blackwater Falls**. This 62-foot cascade is impressive during high water. The naturally occurring dark color of the water is due to the leaching of tannins from the local vegetation and soils.

For more information of bird watching opportunities in the Canaan Valley area contact: John Northeimer, P.O. Box 686, Davis, WV 26260.

Dolly Sods and Bear Rock
Reba and Gene Hutton

Traveling to **Dolly Sods** is not easy! FSR 75 and 19 are gravelled, narrow and rough. Heavy traffic may be encountered on weekends.

From Petersburg, follow SR 55 west and SR 28 south past Cabins to Jordan Run Road. Turn right on to Jordan Run Road (CR 28/7) and proceed to FSR 19 and the sign to Dolly Sods. Turn left onto the gravel road and continue to the top of the mountain. At the top, FSR 19 turns left and leads to the DNR cabin at the bottom of the mountain. Turn right on FSR 75 and immediately find the information board for the Dolly Sods area.

From Canaan Valley, follow SR 32 south to Laneville Road (CR 45). Turn left and follow the Laneville Road about 10 miles. Just past the bridge over Red Creek, you will find the DNR cabin on the left, an information board and the beginning of FSR 19.

From Harmon, follow SR 32 north to Bonner Mountain Road (CR 32/3). Turn right and follow it to CR 45, turning right to the DNR cabin.

Bird as you climb the mountain from the DNR cabin (FSR 19), stopping on the right after 2.7 miles. The trail on the right goes out into a marshy area where you may find Winter Wrens, Northern Waterthrushes, Dark–eyed Juncos, Magnolia Warblers, Hermit Thrushes, Veerys and Chipping Sparrows during June, as well as goldthread, trailing arbutus and wintergreen.

Proceed to the FSR 19/75 intersection and drive out FSR 75 towards **Bear Rocks**. You will see rocky plains, upland bogs and sweeping views. The elevation ranges from 2,600 feet to over 4,000 feet. Stop and walk the many trails. At 2.2 miles, you will come to the fire tower lookout. Another 2.3 miles brings you to the **Northland Loop Nature Trail** and another 0.5 miles to the Red Creek Campground. On the right beyond the parking lot, is the location of the **Allegheny Front Banding Station** that operates from August 15 to the end of September. Alder Flycatchers have been heard and seen across the road from the parking lot. Another 0.8 miles along FSR 75 will bring you to an excellent botany area (the open area to the right where a culvert passes under the road).

Some of the birds you can expect to see and hear are the following: Chestnut–sided, Magnolia, Black–throated Blue, Black–throated Green, Yellow–rumped, Ovenbird, Northern Waterthrush, Common Yellowthroat and Canada warblers, Common Raven, Ruffed Grouse, Least Flycatcher, Eastern Phoebe, Black–capped Chickadee, Winter Wren, Golden–crowned Kinglet, Veery, Hermit and Wood thrush, Brown Thrasher, Eastern Towhee and Solitary Vireo. By walking some of the trails, you will hear the beautiful song of the Hermit Thrush and Winter Wren.

Unusual plant communities are one of the attractions of the Sods. Sphagnum bogs, groves of wind–stunted, one-sided red spruce, heath barrens, rhododendron and laurel thickets, provide an endless variety of vistas. Some of the following noteworthy plants may be found in the botany area mentioned above: oceanorus (*Zigadenus leimanthoides*), goldthread (*Coptis groenlandica*), sundew (*Drosera rotundifolia*), three–toothed cinquefoil (*Potentilla tridentata*), gay wings (*Polygala paucifolia*), purple virgin's bower (*Clematis verticillaris*), star–violet (*Dalibarda repens*), and yellow bartonia (*Bartonia virginica*).

By continuing further, you will come to Bear Rocks where you can bird and botanize the Sods or picnic on the rocks that overlook the valley below. If you follow FSR 75, it will bring you out on Jordan Run Road(CR 28/7). Turn right and return to SR 55.

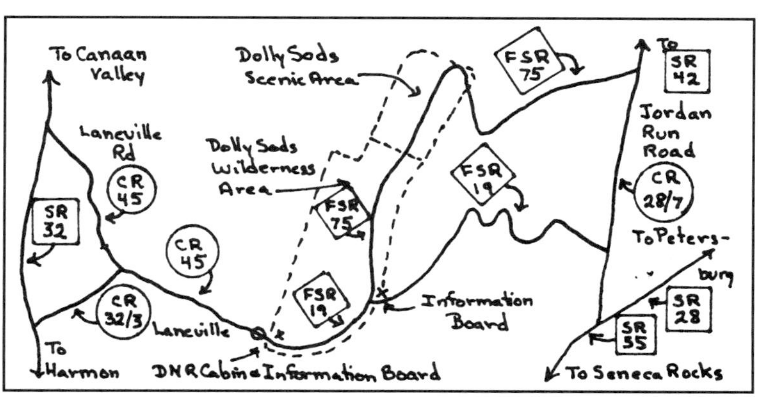

Bickle Knob Recreation Area and Stuart's Park Area

Reba and Gene Hutton

From Elkins at US 250 and US 33, take US 33 east 3.4 miles. Turn left on old Route 33 and go 1.9 miles and then turn left to **Stuart Park** entrance. Go down to the picnic area along the river and listen for Baltimore Orioles, Pileated Woodpeckers and Belted Kingfishers. Return to the park entrance and go up FSR 911 towards **Bickle Knob**. Drive 0.5 mile and park. You should hear and see Wood Thrushes and these warblers: Black and White, Black–throated Blue, and Hooded. Next stop is a gated meadow on your left. Here you will find Eastern Towhees, Eastern Bluebirds, Bobolinks and American Goldfinches. Next, stop at a gated area on your right — here you will hear House Wrens, Tree Swallows, Song and Chipping sparrows, Kingbirds and Brown Thrashers. Listen for Scarlet Tanagers, Rose–breasted Grosbeaks and Common Yellowthroats in the trees across the road.

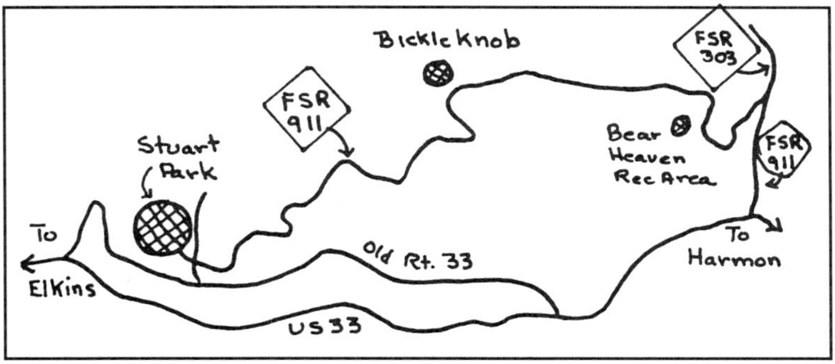

As you proceed up the mountain listen for Red–eyed and Solitary vireo, Dark–eyed Junco, Magnolia and Chestnut–sided warblers. From the tower proceed along the ridge towards **Bear Heaven Recreation Area**. In this area listen and watch for Common Raven, Barred Owl, Red–shouldered Hawk, Winter Wren, Canada and Blackburnian warblers. As you proceed on down the mountain listen on your left for Hermit Thrush and Veery.

When you reach the intersection of 303 go down to the end of the road. Park and walk along the trail. If you are lucky you will hear the Northern Waterthrush. Return to FSR 911 and proceed along the road toward US 33. You will hear many of the above-mentioned species. Watch along the roads for Black Bear, Wild Turkey and Ruffed Grouse.

All of the above species were found during late May and early June 1995, with the exception of the Northern Waterthrush, which was heard in previous years.

If you are interested in botany, you will see beautiful displays of wildflowers. Two rare plants found in this area are *Rubus tomentosa* and *Trillium nivale*. At mile 6.3, look on your right for stair step and knight's plume moss.

Spruce Knob Area

Greg Eddy

Spruce Knob at 4,861 feet is the highest point in West Virginia. From Seneca Rocks, proceed south on US 33 for 9.7 miles. Turn right onto a gravel road, (CR 33/4) which becomes FSR 112. After 7.6 miles, bear right on FSR 104 for 2.5 miles to the top. From Franklin, proceed west on US 33 for about 13 miles to Judy Gap, continuing on US 33 for 1.1 miles before making a left turn onto CR 33/4. From Bartow at the intersection of US 250 and SR 92, proceed north on SR 28 for 10 to 15 miles until you arrive at FSR 112 on the left just over the top to the ridge. FSR 112 will take you to the top of Spruce Knob.

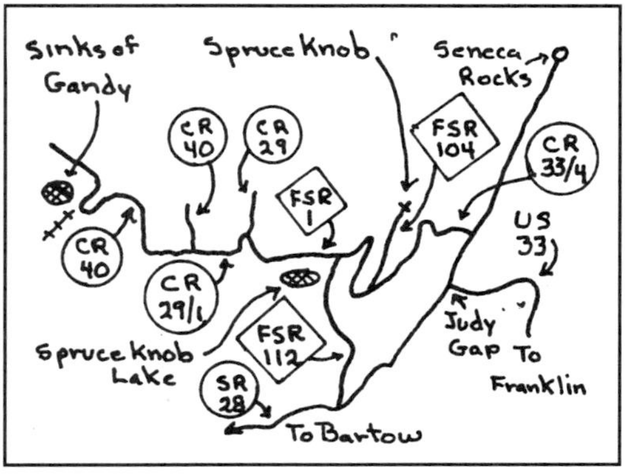

This part of West Virginia contains many species typical of high elevations. Only a little effort is necessary to locate Magnolia, Black–throated Green and Blackburnian warblers, Hermit Thrush, Veery, Winter Wren, Golden–crowned Kinglet, Bobolink, Vesper and Savannah sparrows and Dark–eyed Juncos.

Starting at the parking lot at Spruce Knob look for Yellow–rumped Warblers. This was the first place they were confirmed as breeding in West Virginia. Magnolia Warblers and Dark–eyed Juncos are common here. As the road leaves the parking lot and starts down the mountain it passes through a guaranteed Mourning Warbler location. They can be heard in the treeless area below the road. Dark–eyed Juncos and Eastern Towhees are the most common species here, but Mourning Warblers have always been found. Continue down the mountain to the intersection of FSR 112/104 and bear right.

Turn right at FSR 1 and proceed to **Spruce Knob Lake Recreation Area**. Wood Ducks can sometimes be found on the lake, but they are more common on the beaver ponds found throughout the area. The only Swamp Sparrow found in the area was located along the road just below the dam of the lake. Buckbean (*Menyanthes trifoliata*) can be found along the north side of the lake.

Continuing on FSR 1, bear sharply to the left onto CR 29/1 and left again onto CR 40 and continue to the **Sinks of Gandy**, stopping at a sharp hairpin turn to the right as you proceed up a hill. A rather large spruce bog is located here and contains several interesting plants and birds. The bog is on private land and part of a cow pasture, so show respect for the land owner. Northern Waterthrush are usually found in the bog. Blackpoll and Nashville warblers have been found here in June, but were probably not breeding, although Nashvilles do breed in similar areas of West Virginia. Twinflower (*Linnaea borealis*) can be found about 100 yards along the fence on the right in the bog.

Return to your car and continue another 100 yards and turn left around the corner of the cow pasture. Rather extensive pasture fields are divided by this road. This is about the only location in this area for grassland species including Bobolink, Vesper, Savannah and Grasshopper sparrows. By stopping at the gate on the left, you can climb over the gate and walk along the road until you see the entrance of the Sinks of Gandy on the left where Glady

Creek goes underground. A rather large colony of Barn Swallows is located in the entrance to the sinks. When the water is not too high, it is possible to walk (about one mile) through to where Glady Creek resumes its above–ground course.

The commonest birds that can be seen or heard from the roads in this part of the state during summer in descending order are: Red–eyed Vireo, American Robin, Indigo Bunting, Eastern Towhee, Magnolia Warbler, Veery, Black–capped Chickadee, American Crow, Dark–eyed Junco, Ovenbird, Black–throated Green Warbler, Wood Thrush, Chestnut–sided Warbler and Rose–breasted Grosbeak. You can continue past the Laurel Fork campground to Glady via FSR 1 and CR 22, then turn right on CR 27 north to Alpena on US 33, or you can retrace your steps and return the way you came.

Gaudineer and Barton Knob

Greg Eddy

One of the most interesting birding spots in West Virginia and indeed the Central Appalachians is the **Gaudineer Knob**. This is one of the best places in the state to observe several northern species including; Northern Waterthrush, Golden–crowned Kinglet, Red Crossbill and Yellow–rumped Warbler. Several other high elevation species that are found throughout this part of the state are common here. These include Blackburnian, Black–throated

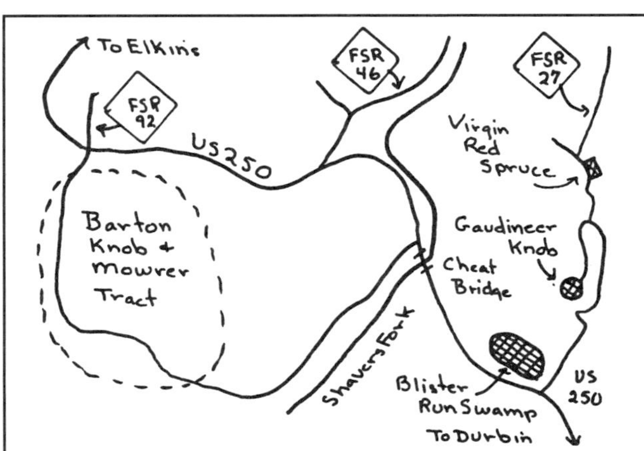

Blue, and Black–throated Green warblers, and Dark–eyed Juncos. In his book *The Appalachians*, Maurice Brooks points out that Gaudineer Knob area is one of the few places in the country that four species of brown–backed thrushes can be heard, the Wood, Hermit, and Swainson's thrushes and the Veery. He also suggests that visitors to the area must divide their time between different fields of nature study. The area is as famous for its numerous species of plants and salamanders as it is for its bird species.

If starting from Elkins follow US 250 south. Sixteen miles south of Elkins, at Huttonsville, US 250 turns to the southeast and climbs Cheat Mountain. A dirt road, FSR 227, leading south from the top of Cheat Mountain gives access to Barton Knob and the **Mowrer Tract**, a very large area south of US 250 that has been strip mined and reclaimed. These reclaimed areas are large grassy flats and therefore perfect for grassland birds. These include Bobolinks, and Vesper, Savannah and Grasshopper sparrows. A near maze of roads exist in this area, but if you carefully follow the main one for about 13 miles, it crosses Shavers Fork of the Cheat River and joins US 250 just east of the river. This general area is called **Cheat Bridge**. When mining and timbering were more active this area was built up but now mostly only remains of buildings can be seen. The Cheat Mountain Lodge, a privately run trout fishing lodge, is also here. One may botanize at the Cheat Bridge or proceed north on US 250 and make a right on FSR 46 after 1.9 miles to botanize along this road and Shavers Fork. Then return to the Cheat Bridge.

Only 1.3 miles after crossing Cheat River US 250 reaches the top of Shavers Mountain. This point is about 3.5 miles west of Durbin. A few hundred yards west of the top of the mountain FSR 27 intersects US 250. Turn north on to this gravel road and stop as soon as it enters the woods. This is one of the best places to hear a Swainson's Thrush. On a June or early July evening Hermit and Swainson's thrushes can both be found here. A **swamp on Blister Run**, a few hundred feet to the left, contains several nesting high elevation species, including Brown Creeper, Golden–crowned Kinglet, Magnolia and Canada warbler, Red–breasted Nuthatch and Northern Waterthrush. US 250 parallels the swamp so that most of its birds can be heard from the paved road.

About a mile from US 250 the FSR has a branch to the left. The left branch ends after 0.5 miles at the top of **Gaudineer Knob**. A very nice picnic area and overlook are maintained here by the US Forest Service. Winter Wrens and Magnolia Warblers are common species here, but Dark–eyed Juncos and Golden–crowned Kinglets are the most numerous. Yellow–rumped Warblers are often found here during the mating season and are probably breeding.

Backtrack and continue less than a mile on the FSR to a small parking area on the right. A maintained trail starts here and circles through a stand of virgin red spruce. Blackburnian Warblers and Golden–crowned Kinglets are two of the commonest species in the spruce. The trees are so tall that the birds are difficult to see, but can be easily heard. Although this is one of the best places in the state to see Red Crossbills, they cannot be guaranteed. They are nomadic and very unpredictable. They are to be looked for in the spruce groves throughout this area. Brown Creepers can be found throughout most of the area, but the best place to see an active nest is where the road passes a swamp only a short distance past the virgin spruce.

Many surprises still remain to be discovered in this part of the state. In recent years several northern species have been found during the breeding season and may be breeding here. These include Northern Goshawk and Olive–sided Flycatchers. White–throated Sparrows have been found breeding and Yellow–rumped Warblers are almost surely breeding here as they are a short distance away at Spruce Knob.

There is good botanizing in Blister Run Swamp.

Cranberry Glades and Scenic Highway
George Breiding

One of the state's prime sections for high–country birding includes the **Cranberry Glades Botanical Area** and the **Highland Scenic Highway**, sometimes referred to as "by–way" or "parkway." Parts of those areas are closely connected on the **Monongahela National Forest** in southeastern Pocahontas County.

The focal point for "working" them is the **Cranberry Mountain Visitors' Center** located at the intersection of SR 39/55 and SR 150. To reach the visitors' center from Marlinton, proceed south on US 219 and SR 55 about eight miles to Mill Point. Bear right and continue on SR 55 for seven miles to the visitors' center. From Richwood proceed east on SR 39/55 to the visitors' center. From Lewisburg, go north on US 219 to Mill Point and bear left on SR 55. One can begin at the northern entrance of the Scenic Highway by driving from Marlinton 7 miles north on US 219 to the intersection with SR 150.

Cranberry Glades — The glades are one–half mile west of the center on SR 39 and SR 55. The elevation of the floor of the glades, an upland acid bog, and often referred to as "the muskeg furthest south," is approximately 3300 feet above sea level and embraces 750 acres of five glades. They gained their reputation for their flora, namely the terrestrial orchids: the rose pogonia or snake's mouth, *Pogonia ophioglossoides*, grass pink, *Calapogon pulchellus*, and purple fringed orchid, *Platanthera grandiflora* , which mostly bloom in late June and early July.

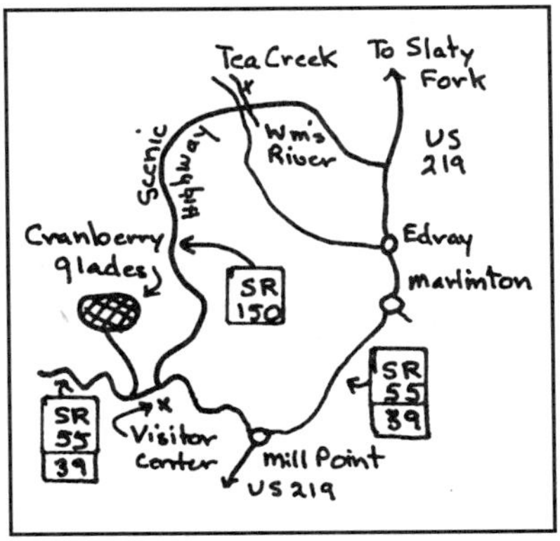

Only parts of two of the glades are accessible to the public. A one–half mile **boardwalk** leads through several plant communities where you may encounter (especially in the month of June) the Canada and Magnolia warblers, Northern Waterthrush, Alder Flycatcher, Swamp Sparrow, Purple Finch, Red–shouldered Hawk, and Common Raven. The Olive–sided Flycatcher occurs sporadically. In the general environs look for Black–throated Blue, Mourning, and Parula warblers, Red–breasted Nuthatch, Veery, and Hermit and Wood thrushes.

Some of the plants read like a "who's who" of species found in some of the states hundreds of miles northward and lower Canada, such as bog rosemary *(Andromeda glaucophylla*), dwarf dogwood, Canada mayflower, chokeberry, wild raisin, buckbean, *Menyanthes*, and yew. The roundleaf sundew is readily observed along the boardwalk, as are two species of cranberry, and the pitcher plant, which has been introduced.

The yellow–necked vole, a disjunct small mammal of more northerly regions is known to occur. Black bears are frequently seen on and from the boardwalk. Conducted nature walks are scheduled on weekend afternoons during the summer season. While at the glades consider walking the **Cow Pasture Trail** and the road to the **Back Country area**. The first is easily reached on foot from the glades parking lot while the second is reached by driving on to the next parking lot.

Scenic Highway — The scenic highway traverses 22.5 miles of exhilarating roller–coaster panorama of 3150 to 4200 feet above sea level with some of the surrounding ridges rising to about 4700 feet. The highway is an experience, not only for birders, but for botanists who will find the flora of the islands of the boreal spruce forest and the northern hardwoods (beech–birch–maple) varied and challenging.

Some of the species of birds along the highway are similar to those found in the glades, but the road is a wide–open place to roam. There is very little traffic and you can walk to your heart's content as you hear the long, bubbly, tinkling repertoire of the Winter Wren. Many opportunities will arise to listen to and try to find the Mourning Warbler, sometimes common in the **Williams River area**. A road beneath the bridge leads to **Tea Creek Campground** where you are likely to hear the Veery and in some years, the Wood Thrush, both primarily deciduous woodland species.

On some of the roads leading off the highway, there's a chance you may see deer, coyote, and Wild Turkey. A black bear sanctuary has been established along the highway, and a road leads through it. Sometimes it's gated to vehicular traffic, but offers an excellent place to hike in solitude and look for birds. One June day, a Swainson's Thrush was heard on one side of the logging road and a Hermit on the other. These two thrushes occur in the spruce belt and surrounding deciduous forest in the higher elevation. If you slowly cruise and make many stops along the highway between the Williams River bridge and the Cranberry Glades overlook parking lot, and weather conditions are calm on a mid–June evening, you may hear all four of the "brown" thrushes within proximity of each other. Some years the Wood Thrush is absent.

The chickadee you hear and see most likely will be the Black–capped Chickadee. The Indigo Bunting, Chestnut–sided and Mourning warblers are among the most common species along the road. Cutting through a virtual wilderness has created habitat resulting in those species becoming communal neighbors.

There are overlooks with panoramic views, some with picnic shelters and toilets. At the **Big Spruce overlook**, there's a short boardwalk trail where there are examples of the Fraser fir. In the triangle and along the marshy areas at the parking lot, you can find twayblade, adder's tongue fern, grape fern and clubmoss. Stop and spend time hiking in the areas of several of the overlooks. You might add to your list: Slate–colored Junco, Black–throated Green, Black and White and Blackburnian warblers, Scarlet Tanager, Rose–breasted Grosbeak, Eastern Towhee, Red–breasted Nuthatch, Golden–crowned Kinglet, Cedar Waxwing, and Common Raven. With a bit of luck, you may see wandering flocks of Red Crossbills, perhaps a Bald Eagle in late summer or hear a Great Horned Owl in late afternoon. Butterfly chasers should be on the lookout for the Pipevine Swallowtail. The Dutchman's pipe or pipevine occurs throughout the area.

The forest is laced with a network of hiking trails, camping areas and opportunities for mountain biking. At the visitors center you can find out about special places of scenic interest plus the state parks and forests in the vicinity. All of them are places to look for birds.

Be prepared for all kinds of weather. Wear sturdy footwear if you plan to hike some of the trails. They can be rocky, steep, and muddy. Rain gear and warm garments may be needed on short notice. Have drinking water available or a container to get some at the visitors' center.

For information about the glades or highway contact: (1) Cranberry Mountain Visitors Center, USDA Forest Service, P.O. Box 110, Richwood, WV, 26261; (2) Gauley Ranger District, USDA Forest Service, P.O. Box 110, Richwood, WV, 26261, (304) 846–2696; and (3) Marlinton Ranger District, USDA Forest Service, P.O. Box 210, Marlinton, WV, 24954,

(304) 799–4334. The visitor center is open Memorial Day through Labor Day, 9:00 to 5:00 daily; Jan. through April & Nov. Sat & Sun. 10:00–4:00; May, Sept, & Oct, Fri, Sat, Sun. 9:00–5:00; closed December.

For lodging, restaurants and other points of interest contact: (4) Pocahontas County Tourism Commission, P.O. Box 275, Marlinton, WV, 24954, 1–800–336–7009; and (5) Richwood Area Chamber of Commerce, 50 Oakford Avenue, Richwood, WV 26261, (304) 846–6790.

If you are an out–of–state visitor, enjoy the state and some of the areas that John Denver may have had in mind when he wrote and recorded "Almost Heaven."

Snowshoe Moutain Resort

Karen McClure

Birding at Snowshoe is an unique opportunity to observe 11,000 acres with an elevation drop of over 1500 feet and the diversity of species that go with the change in altitude. The resort is on top of one of the highest peaks in the state, 4848 feet. Pine Siskins are easily observed at the very top of the mountain and Killdeers, Ruby–throated Hummingbirds, Blue Jays, Common Ravens, American Robins, Cedar Waxwings, and Dark–eyed Juncos are abundant in the park–like resort setting. Hiking maps are available at the Snowshoe Adventure Center. Trails to Shaver's Lake are strenuous, but worth the effort for the avid birder. Take one of the ski/hiking trails down to Shaver's Lake from the top and you may find the following: Red–tailed Hawk, Ruffed Grouse, Wild Turkey, Downy Woodpecker, Eastern Wood–Peewee, Barn Swallow, Black–capped Chickadee, Tufted Titmouse, Red–breasted Nuthatch, Carolina and Winter wren, Hermit and Wood thrushes, Veery, Gray Catbird, Solitary and Red–eyed vireos, Blue–winged, Magnolia, Black–throated Green, and Chestnut–sided warblers, Ovenbird, Common Yellowthroat, Scarlet Tanager, Northern Cardinal, Rose–breasted Grosbeak, Indigo Bunting, Eastern Towhee, Red Crossbill, Chipping and Song sparrows, Eastern Meadowlark, Northern Oriole, and American Goldfinch.

Snowshoe offers a variety of lodging choices, restaurants, and a full program of nature and recreational programs. For more information, call (304) 572–1000.

3. Mountain Lakes Region
Joey Herron

The Mountain Lakes Region is located in the center of West Virginia, with I-79 running north–south through the area. Some of the state's best lakes are located here. **Stonewall Jackson Lake, Stonecoal Lake, Burnsville Lake**, and **Sutton Lake** are all just off I–79 with easy access. During the spring and fall migrations these areas are active with passing waterfowl and other migrants. Holly River State Park has exceptional wild flowers in the spring.

With easy access from the Interstate, many excellent birding opportunities exist within this region. Breeding Blue–winged and Golden–winged warblers are active in many areas together, along with Hooded and Kentucky warblers. Many farms in the area yield numerous grassland species including Indigo Buntings, Field and Savannah sparrows, Meadowlarks, American Kestrels and Red–tailed Hawks. Wooded areas along the Interstate and the lakes are prime for Scarlet and Summer tanagers, over twenty species of warblers, Wood Thrushes, Mockingbirds, and Brown Thrashers.

During the breeding season from May to July over seventy species can be observed in just a few hours. This area is just two hours south of Pittsburgh and four and one half hours west of Washington, D.C. The Mountain Lakes Region has some of the best and most exciting birding in West Virginia.

Jackson's Mill
Joey Herron

Jackson's Mill is located in north central West Virginia, easily accessible from I–79. Take either exit 99 (Weston/ Buckhannon) or exit 105 (Jane Lew). From exit 99, go to Weston and follow US 19 north to Jackson's Mill. From exit 105, go to Jane Lew, turn left on US 19 and go approximately

6 miles south to the New Haven Motel and turn right. Follow CR 12 to the main entrance of Jackson's Mill.

The Mill has both birding and historical significance. It was the boyhood home of General Thomas "Stonewall" Jackson, the famous Confederate General of the Civil War. Besides the birding, take time to visit the museum and gift shop on the grounds, as well as the restored mill. Overnight accommodations are available at the lodge or in dorm–type houses. The dining hall has wonderful homemade meals served family style. It is the State 4–H Camp and Conference Center. Be sure and pick up a birding checklist at the registration counter or the gift shop.

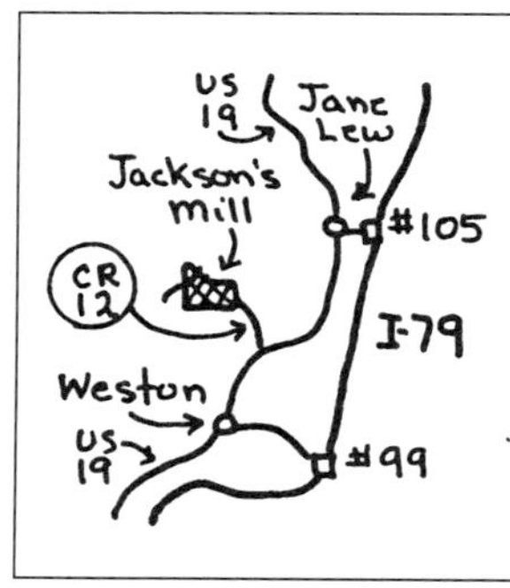

Diverse habitat surrounds Jackson's Mill, making it an interesting area for a wide variety of species. The **West Fork River** runs through the property and a small air strip provides a large open field area for a number of species. Pileated and Red–headed woodpeckers and Yellow–throated Warblers are active along the river. Parula Warblers nest on the grounds along with a family of Barred Owls. Louisiana Waterthrushes and Wood Thrushes nest in the large stand of woods behind the mill. Over fifteen species of warblers can be seen within walking distance of the main grounds. Across the river and over the old railroad bed, approximately two hundred meters (600 feet) from the main entrance bridge, is a gate that leads the way up to **"State Hollow"**. Many woodland species can be observed, including Baltimore Orioles, Scarlet Tanagers, Cerulean and Kentucky warblers, and Wood Ducks that usually hang around a small pond at the head of the hollow. The **airstrip** has had Bobolinks, Vesper Sparrows and Meadowlarks. A Kestrel or Red–tailed Hawk occasionally shows up to hunt the open area for grasshoppers and rodents. One year, twenty Bonaparte's Gulls were on the landing strip, probably thinking it looked like water from their high vantage point when flying over. They stayed for a whole afternoon and later moved on to Stonecoal Lake, another good birding area to the southeast.

Another migratory bird was the Red Crossbill.

Over one hundred species have been recorded at Jackson's Mill, including twenty species of warblers, seven species of woodpeckers and four vireo species. Both Summer and Scarlet tanagers can be seen in the same day. Also Baltimore and Orchard orioles are common in the area. Yellow–throated Warblers frequent the sycamore trees along the river and Parula Warblers flutter among the pines.

For more information on the Lewis County and Jackson's Mill area, contact the Lewis County Convention and Visitors Bureau in Weston at 134 Center Ave, P. O. Box 379, Weston WV, 26452, (304) 269–7328 or Jackson's Mill, PO Box 670, Weston, WV, 26452, (304) 269–5100.

Bird List for Jackson's Mill

Great Blue Heron
Green Heron
Black Duck
Wood Duck
Sharp-shinned Hawk
Broad-winged Hawk
Red-shouldered Hawk
Red-tailed Hawk
American Kestrel
Northern Bobwhite
Solitary Sandpiper
Spotted Sandpiper
Least Sandpiper
Pectoral Sandpiper
Yellow-billed Cuckoo
Kentucky Warbler
Eastern Screech-owl
Barred Owl
Belted Kingfisher
Eastern Bluebird
Swainson's Thrush
Hermit Thrush
Wood Thrush
Mockingbird
White-eyed Vireo
Yellow-throated Warbler
Warbling Vireo
Blue-winged Warbler
Yellow-throated Warbler
Palm Warbler
Cerulean Warbler
Worm-eating Warbler
Louisiana Waterthrush
Hooded Warbler
Summer Tanager
Indigo Bunting
Orchard Oriole

Stonecoal Lake

Joey Herron

Stonecoal Lake stretches four miles over Lewis and Upshur counties, and is surrounded by over four thousand acres of prime birding habitat. The west end of the lake can be approached by the **Georgetown Road**, which cuts off US 33 at Horner by taking CR 15 south. The east end of the lake is accessible via the **Bushy Fork Road** (CR 7), which is just west of Buckhannon on US 33. The west end of the lake is approximately four miles from exit 99 on I–79 via US 33, and the east end is approximately 12 miles from exit 99 on I–79.

The upper end of the five hundred acre lake is a prime stopover during migration for hundreds of waterfowl and shorebirds. When the water is low, a huge mud flat is exposed and creates a perfect area for ducks and shorebirds to feed. Greater Yellowlegs, Least Sandpipers, Semipalmated Plovers, along with Ruddy Ducks, Oldsquaws, and Scaups, all frequent this area. It is preferable to have a scope to identify the birds along the shore, although binoculars will do the job quite well. Be sure to watch for a pair of Red–tailed Hawks that hunt this area all year long.

Tundra Swans, Bonaparte's Gulls, Common Loon, Bufflehead and three species of mergansers have been noted near the lower end of the lake which is easily accessible from US 33. There are three or four places along the Georgetown Road, which runs high above the lake, to see half the lake surface and observe any waterfowl that might be present. There are trails and boat ramps all along the lower end, giving easy access for hiking or boating.

The fields surrounding the lake are active with Grasshopper Sparrows, Meadowlarks, Red–winged Blackbirds, Bluebirds, and Killdeer. A number of warblers can be seen in many of the wooded areas along the shores and bordering the fields. An abundance of wildlife

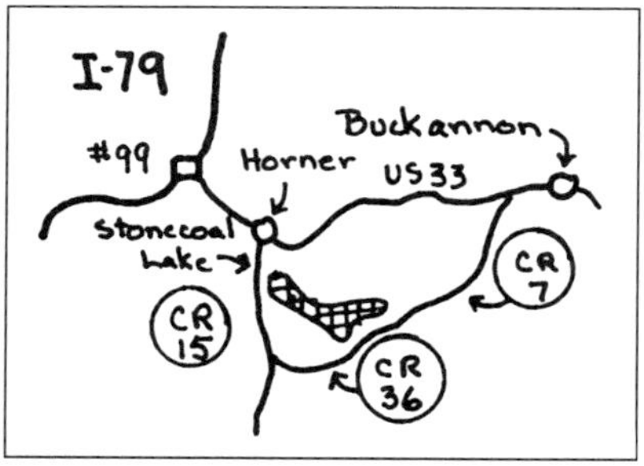

populates the area including deer, beaver and squirrels. Both early spring and early fall are excellent times for birding.

For more information concerning Stonecoal Lake, contact the Lewis County Convention & Visitors' Bureau in Weston, West Virginia. They can provide a map and directions to the lake.

Stonewall Jackson Lake

J. Lawrence Smith and James F. Triplett

The creation of **Stonewall Jackson Lake** by the impounding of the West Fork River south of Weston, Lewis County, in 1985 brought into being wetlands and riparian habitat that greatly increased the variety of birdlife in this area of central West Virginia. The drainage area above the dam covers 100 square miles with 82 miles of shoreline when the water is at summer level. The lake has roughly the shape of an inverted "V," and along with the main West Fork, impounded upstream to near Walkersville, has two sizable arms. Skin Creek and its upper branch of Little Skin Creek form the larger, while Sand Fork forms the smaller.

Physiographically, the lake is located on the ancient tableland of the Allegheny Plateau of the Appalachian Province. The underlying strata of sandstone, shale, and some coal are of Paleozoic age and date from the late Pennsylvanian and early Permian periods. The elevation at the dam is about 1000 feet above sea level, while the crests of some ridges in the upper area of the lake approach 2000 feet.

The forested slopes are characteristic of a Central Hardwoods type, with mixed oak–hickory on the elevated ridges. A more mesic mixture of hardwoods such as American beech, sugar maple, and yellow poplar is found on lower slopes and in shaded ravines. Common species in riparian situations and on the flood plain are silver maple, black willow and sycamore.

A stop at the dam is convenient for the birder proceeding south on I–79. Begin a visit to the lake with a stop at the **visitor center** at the dam, located about 3 miles south of

Weston. Take the South Weston exit 96 off I–79, turn left onto CR 30, and follow this road for a short distance until a right turn leads to the dam. The visitor center, a facility operated by the Army Corps of Engineers, offers displays of items relating to local history, the construction of the dam, and an excellent habitat group of local wildlife.

Birders traveling north will find it easier to make the dam the last stop after traveling the following route around the lake: Take the Roanoke exit 91 off I–79 and follow US 19 south 2.7 miles to the entrance of Stonewall Jackson Lake State Park. Continuing another 0.8 miles on US 19, turn right on **Oil Creek Road** and park near the bridge. This is the first stop on a 25-mile route around the lake for stops at the more productive birding spots along the way.

During the breeding season a number of species can be expected along the brushy slope across from the bridge and in taller trees along the shore. Birds recorded here are Yellow Warbler, Yellow–breasted Chat, White–eyed Vireo, Prairie Warbler, Common Yellowthroat, Eastern Towhee, Field Sparrow, Golden–winged Warbler, and Orchard Oriole. Willow Flycatchers have been found in this area, and a small colony of Cliff Swallows nests beneath the bridge.

A view of a large area of the lake is available from the vicinity of the bridge and the water should be scanned for waterfowl in season. Another especially good view of the lake is reached by continuing along Oil Creek Road for 1.9 miles.

At the intersection of Oil Creek Road and the **Walkersville–Arnold Road** (CR 44), proceed toward Walkersville. (This intersection is hazardous. Use caution when turning left.) The road leads through a variety of habitats such as open field, brushland and woods. Yellow–breasted Chats have been found in numbers in the roadside thickets, and Hooded, Cerulean and Yellow–throated warblers may be heard in the woods. Follow the Walkersville–Arnold Road for 2.1 miles to the bridge at what was once the site of Jacksonville. There is much standing dead timber here and elsewhere along the lake edge, which provides cavities for nesting Tree Swallows. Wood Ducks have been found in numbers upon their appearance in early March and on occasion other ducks have been rather abundant in early spring. Among waterfowl found here are Ring–necked Ducks, American Wigeon, Lesser Scaups, Gadwalls, Canvasbacks, Northern Pintails, Hooded and Red–breasted mergansers.

When the water level is low and mud flats are exposed, shorebirds can be expected. A good variety of sandpipers has been found in spring and fall, including Greater and Lesser yellowlegs, and Solitary, Spotted, Semipalmated, Least and Pectoral sandpipers. Among birds less frequently recorded and rather unexpected have been Sanderlings and Dunlins.

The road beyond the bridge at Jacksonville continues along the winding course of the upper lake for 1.5 miles to an intersection with **Abrams Run Road**. Red–tailed Hawks can be expected in the area in March and Ospreys will be seen later in spring and have remained to breed at the lake in recent years.

A walk along the railroad which crosses the Abrams

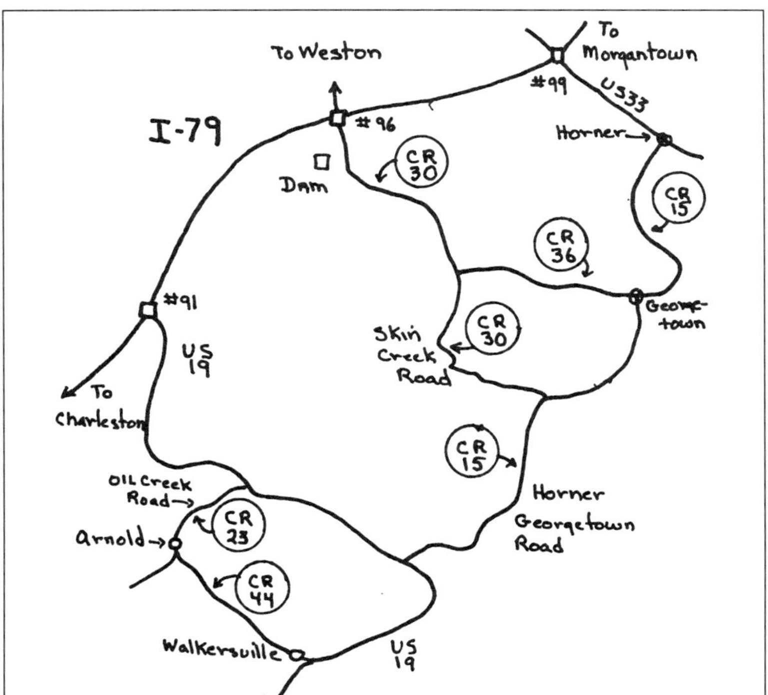

Run Road offers access to bays where waterfowl can be expected in season and herons such as the Great Egret have been found. A low–lying field, a distance of 0.5 miles along Abrams Run Road, becomes flooded in spring and often proves productive for waterfowl. Wood Ducks, Mallards, and Blue–winged Teals are often found in numbers. Continue on the Walkersville–Arnold Road from the intersection with Abrams Run Road for 1.6 miles to US 19 at Walkersville. Turn left on US 19 and drive 3.8 miles to the **Horner–Georgetown Road** (CR 15). Wild Turkeys have occasionally been observed feeding in the fields near the highway.

Turn right on the Horner–Georgetown Road and follow it for 2.7 miles to the intersection with **Skin Creek Road** (CR 30) at Vandalia. At 0.5 miles from US 19 there is a marshy field to the right and below the road. In spring Wood Ducks, Blue–winged Teal and other species such as Northern Shovelers have been recorded here.

Waterfowl can be expected in season along the upper lake at **Vandalia**, and Double–crested Cormorants have frequently been found here in numbers. Goslings of the largely resident Canada Geese may be found here in summer along with broods of Mallards and Wood Ducks. American Kestrels seem to be rather permanent in the area.

Turn left on Skin Creek Road which continues through open field and brushy hillsides for 3.7 miles to Little Skin Creek Road which can often be productive for waterfowl in season. At 0.5 miles beyond Little Skin Creek is **Curtis Run**. A right along the road leads past a series of beaver ponds and extensive areas of cattail where Marsh Wrens and marsh birds such as Sora and Virginia rails occur. Woodcock have been heard performing their courtship "sky dance" in the area and Whip–poor–wills have been heard on summer evenings. Owling has proven productive here, as well as elsewhere along the lake, with numerous Barred and Eastern Screech-owls responding to recordings of their voices. Return to Skin Creek Road and continue 3.9 miles to the dam and nearby access to I–79.

4. New River/Greenbrier Valley

Jim Phillips

Southern West Virginia offers birders a variety of ways to enjoy their hobby at different levels of intensity and is easily accessible from the north and south by I–77 and from the east and west by I–64. The truly adventurous folks may enjoy birding the **New River Gorge** from whitewater rafts while other river enthusiasts may be content to paddle canoes along the **Bluestone** or **Greenbrier** rivers or milder stretches of the **New River**. Birders who wish to have no current to move them along may be interested in **Lake Sherwood**, **Moncove Lake** or **Bluestone Lake**. Lake Sherwood is a rather small lake (a little over three miles around) located in the southern portion of the Monongahela National Forest. Facilities include campgrounds, picnic areas and a swimming area. Moncove Lake is a 144 acre reservoir within **Moncove Lake State Park**. Moncove offers camping and picnic areas and a seasonal swimming area. The situation of these two bodies of water, in the mountains, should be inviting to migrating waterfowl but little is in the literature. **Bluestone Lake** is about 2,000 acres in size and includes U.S. Army Corps of Engineers property and the **Bluestone Wildlife Management Area**. Primitive camping is available and there are several access sites.

Land loving birders may also enjoy the area. Hiking trails of various lengths are available in the state parks, state forests, state wildlife management areas, national parks and national forests. Birders interested in longer walks may wish to backpack or dayhike on the **Allegheny Trail**(which will eventually cross West Virginia to connect Pennsylvania and Virginia) or the already complete **Greenbrier River Trail**. The Greenbrier River Trail is about 80 miles in length, passing through or near several state parks between Caldwell (to the south) and Cass (to the north) where you can take a narrow gage train to the top of **Bald Knob** (4,842 feet). This is part of the rails to trails program. Maps and trail guides are available for both of these areas. Less vigorous birding trips may be conducted by taking scenic drives. Routes such as US 219, 60, 19, and 460 and SR 20, 92, 41, 16,

3, and 12 offer not only access to birding sites but birding potential along the way. However, these roads may be busy with nonbirders so be careful. In fact, it is usually advisable to explore the smaller, less busy county roads. Less active, more patient birders may be even less mobile and visit one of the fall hawkwatching sites in this region. Here, the birder may sit on a rock or in a lawn chair and enjoy the migration.

Southern West Virginia offers birders most everything they need. Private, state and federal lodging facilities are available whether you are interested in backpacking, camping, cabins, bed and breakfasts, lodges or hotels. Grocery stores, restaurants, service stations and visitor's centers also provide for your needs while traveling. All you need do is find the birds!

Greenbrier River Trail

Josh Lipton

A dominant natural feature of Summers, Greenbrier and Pocahontas counties is the **Greenbrier River**. Access to the river by boat can be made at any one of several public access points but is possible only when water levels are sufficient. Access via the the Greenbrier River is possible year round and provides access to over 75 miles of riparian habitat. The trail is a converted railroad right of way that is managed by the state. Travel on foot, bicycle, or on horseback are permitted. Some primitive campsites are available along the trail. The trail begins at Caldwell, WV and continues north to Cass,WV. Access to the trail head at Caldwell is from exit 169 off I–64. Drive south on US 219 to Lewisburg which has many craft and antique shops. Turn left on US 60 east and continue towards Caldwell. Turn north on Stonehouse Road (CR 38) to the parking area. Access roads to the trail are well marked along US 219. For a map that locates the access points, contact the Superintendent at Watoga State Park, Marlinton, WV, 24954, 1–304–799–4087.

The Greenbrier River flows through mountainous terrain that is mostly forested. It passes by small towns, summer homes and farms. Access to the river trail at **Renick** provides an excellent introduction to the Greenbrier. This is one of the best places along the river to observe birds and wildlife. The turn off onto Auto Road from US 219 at Renick is well marked. Before reaching the bridge turn left and travel to the north end of Renick where you will find a parking area adjacent to the trail. Following the trail north (upstream) past mile markers 26, 27, and 28 takes you through a variety of habitats including a small wetland area around mile 28. Limestone cliffs border the trail in parts and the cliffs border agricultural land above. A nice variety of habitats along this section of the trail allows for diversity of species that can be seen. In addition to bird species, keep an eye open for muskrat, beaver and river otter.

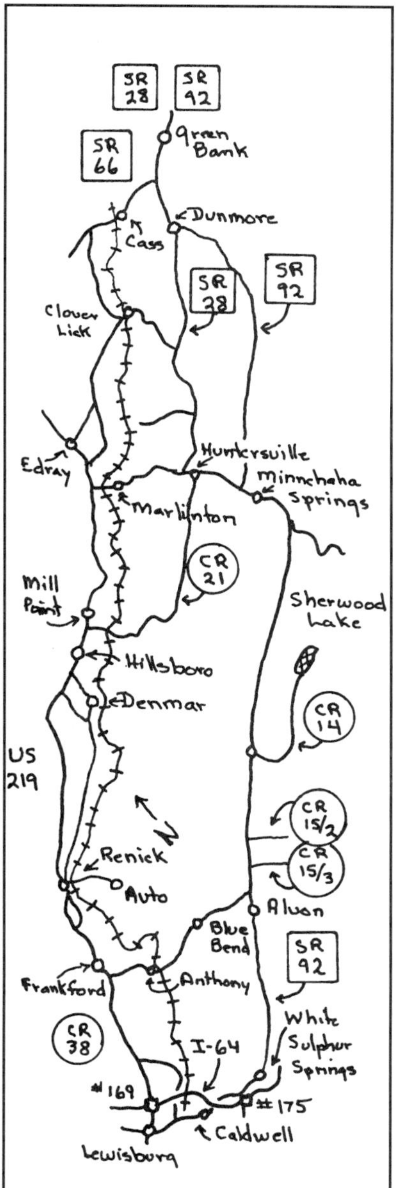

A variety of waterfowl may be observed during migration but they are not abundant. Canada Geese and Wood Ducks breed in the area. Great Blue Herons are regularly seen, especially in winter. Green Herons are summer residents. Spotted Sandpipers are mostly seen in spring. Osprey stop to feed during migration. Great

Horned and Eastern Screech–owls are the owls most regularly encountered along the river. Belted Kingfisher breed along the river with some taking up winter residence. Cliff Swallows sometimes nest on the bridge at Renick. Northern Rough–winged Swallows use kingfisher burrows or crevasses in the cliffs for nesting. Common Ravens utilize the abundance of cliffs along the river for nesting as well. Brown Creeper, Winter Wren and Hermit Thrush are present in winter. Golden–crowned Kinglets can be numerous in winter. Red–eyed, White–eyed and Solitary vireos are all summer residents. Swamp, Fox, and White–crowned sparrows are regularly seen during migration. White–crowned Sparrows can often be found in winter as well. White–throated Sparrows are regular winter residents.

More than 22 species of warblers have been observed in this area at various times. Warblers that are sometimes associated with riparian habitats that are found here are: Northern Parula, Yellow–throated, Blue–winged, Common Yellowthroat, and Louisiana Waterthrush. Louisiana Waterthrush arrive early in April and their song can be heard at almost any point along the river at this time. One of the few places on the river to find Yellow–throated Warblers is here at Renick between mile post 26 and 27. Other warblers include Black–and–white, Worm–eating, American Redstart, Yellow, Chestnut–sided, Ovenbird, Yellow–breasted Chat and Yellow–rumped warblers being present in winter.

Cold Knob
Josh Lipton

Cold Knob with an elevation of 4,345 feet, is the second highest point in Greenbrier County, West Virginia. The highest, at 4,372 feet, is Grassy Knob just to the northwest. Cold Knob seen from a distance is recognized by the high wall cut around its entire circumference, a result of mining. From the peak, views of the valley are spectacular on a clear day. Cold Knob and the area to the north are properties owned by the Westvaco Corporation and limited access to the public is allowed. For more information on birding on Westvaco property contact Mark Ford at (304) 392–1538.

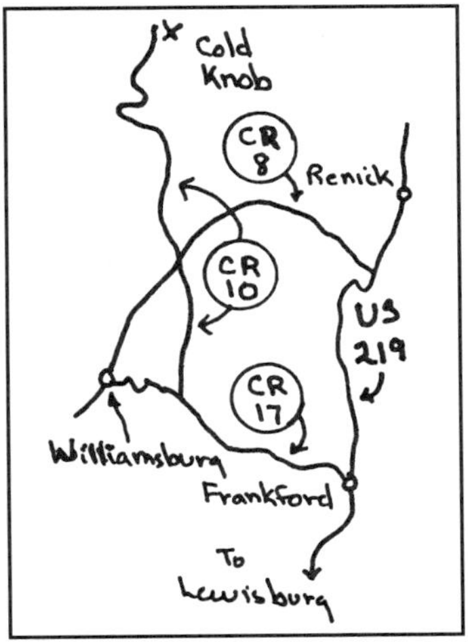

From I–64, take exit 169 and go north on US 219 for approximately 9.5 miles to Frankford. Turn west on CR 17 and travel 7 miles to CR 10. Turn left and take CR 10 north In the first mile, look for Red-headed Woodpeckers in the oaks. Go for 9.75 miles to **Cold Knob Viewing Area** sign and turn right. This takes you to the Cold Knob rim which is a loop road. Return to the Cold Knob sign and travel north along the **Beech Ridge Road** to access some of the surrounding areas. These are rough roads that are not covered for the most part on county maps. Log and coal trucks may be active.

The areas surrounding Cold Knob and extending to the north along Beech Ridge for approximately 10 miles have been clear cut at different times and the forest is at various stages of regeneration. Other areas where mining has occurred have been planted in grasses and some pine. Birds of high elevation, common to forested areas at various stages of regeneration that are found here are: Black–billed and Yellow–billed cuckoos, Hermit Thrush, Golden–winged, Chestnut–sided, Black–throated Green, American Redstart, Ovenbird, Common Yellowthroat, and Mourning warblers, Scarlet Tanager, Rose–breasted Grosbeak, Indigo Bunting and Dark–eyed Juncos. Some unusual birds for Greenbrier County that are found in the high grassland areas are Vesper and Savannah sparrows. Northern Harriers are often found hunting these high meadows during migration.

Greenbrier State Forest

Josh Lipton & Gwen Powell Balogh

The **Greenbrier State Forest** and surrounding area offer

a variety of habitats to explore. In addition to excellent birding some interesting plant communities are found here. From I–64, take exit 175 and turn east onto Harts Run Road, CR/14, for approximately 1.3 miles to the Greenbrier State Forest office where you can pick up a map of the forest. Continue on this road along **Harts Run**, a tributary of the Greenbrier River. There are several trail heads in this area. Eventually the road climbs **Kate's Mountain** and ends on the west side at White Sulphur Springs, directly across from the Greenbrier Hotel.

A special topographical feature of the region are areas referred to as **shale barrens**. These are steep well drained inclines with a surface of loose fragmented shale. The adverse conditions have given rise to a number of plant species endemic to shale barrens in West Virginia. Many of these plants were first found on the Greenbrier State Forest, and the surrounding White Sulphur Springs area. Kate's Mountain Clover, *Trifoluim viginicum*, a shale barren plant, is named for the prominent ridge overlooking the state forest, where this plant was first found. Other shale barren plants that were first found in the area are: Shale Ragwort *Senecio antennariifolius*, Yellow Pimpernel,*Taenidia integerrima*, Shale Evening Primrose, *Oenothera argillicola*, Shale Virgin's–Bower, *Clematis albicoma*, and Yellow Buckwheat, *Eriogonum. alleni*. Look for many other plants unique to shale barrens that can be found here. Shale barrens are fragile habitats so remember to stay on trails and off of steep banks where human traffic could result in the loss of rare plant species.

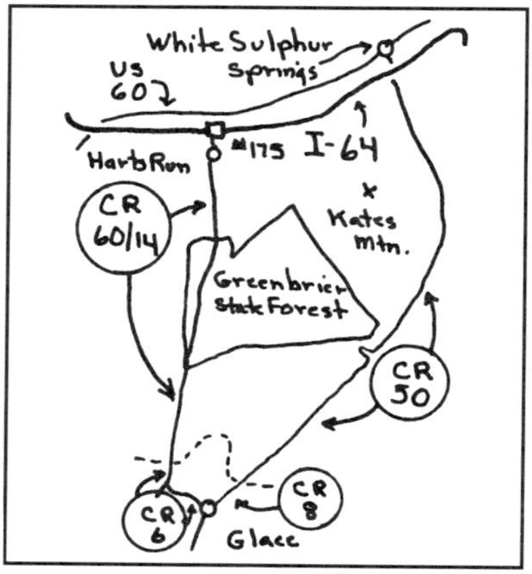

On warm spring and summer days keep your eye open for lizards. The open rocky slopes of shale barrens provide ideal conditions for the Northern Fence Lizard, *Sceloporus*

undulatus..

Low areas of the forest are home to the Acadian Flycatcher, Wood Thrush, Louisiana Waterthrush, Northern Parula, Worm–eating and Black–and–white warbler. Further up the mountain look for Scarlet Tanager, Eastern Wood–Pewee, Yellow–billed Cuckoo, Great Crested Flycatcher, Black–throated Green Warbler and Ovenbird.

Kate's Mountain at the northern edge of the forest is a high ridge with a mix of mature trees making it a favorite place for summer residents such as: Yellow–throated Vireo, Cerulean and Blackburnian warblers. Some adjacent timbered areas also attract Chestnut–sided Warblers and American Redstarts.

Babcock State Park
Emily Grafton

Babcock State Park and the areas just outside its boundaries make for a wonderful place to go bird watching. Dry, upland oak–hickory flats, boulder choked ravines, cascading water, steep hillsides draped with hemlock and rhododendron, beaver ponds, wetlands and wide vistas are the major habitats where a delightful variety of bird life may be seen. From US 19, exit at Hico and proceed east on US 60 through Lookout. Turn south on SR 41 to Babcock SP. Alternately, take SR 61 north out of Beckley and just past Piney View, turn right on SR 41 and proceed to the park after passing through Landisbury.

The **Glade Creek–Manns Creek canyon**, which comprises the major portion of the park's 4,127 acres, is an exciting place to watch for birds year round. Glade Creek cuts across two miles of the park from its eastern boundary, carving a rugged canyon. In an absolutely beautiful setting, Glade Creek flows into Manns Creek, and together they continue a downward plunge for four more miles before joining the New River. The park boundaries end within a quarter mile of the river.

The highest elevation in the park, about 2,700 feet, occurs on the knolls along the southern boundary of the park. The lowest elevation, about 900 feet, occurs where Manns Creek joins the New River. This significant change in elevation within the parks boundaries contributes to the diversity of habitats. Also, the depth of the canyon, coupled with the opposing north/south aspects of its walls create an ideal situation for micro–climates.

A narrow, dirt road traverses the full 6 mile length of the south side of the canyon, from the Administration Building to the river. **Sewell Road**, which is paved for the first mile, was completed in 1878 and took two years to build. For nearly sixty years the road bed provided a base for narrow gauge rail road trains to transport coal from the mines near the cliffs to rows of coke ovens at the New River. At the present time, this serpentine road carries hikers, bikers and fishermen into the otherwise inaccessible canyon and ghost town of Sewell.

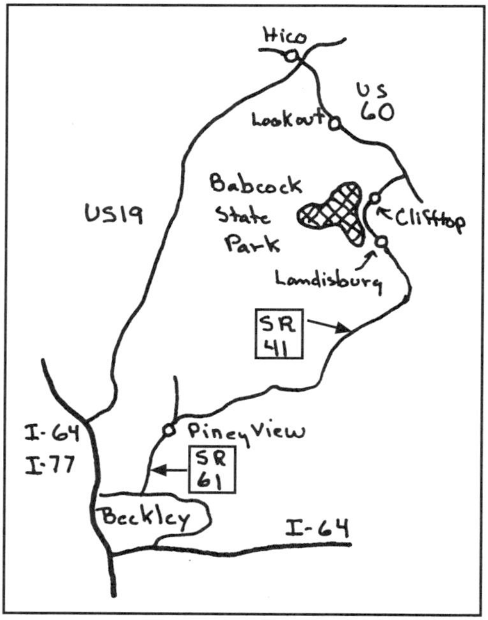

Throughout April and May, thousands of neotropical and other birds migrate north along the New River, venturing up into the canyon to feed on the hatching insects so abundant in the expanding foliage. During the first two weeks of May, it is possible to observe forty – fifty species of birds throughout the canyon and along the river. Blackburnian Warblers may be extremely abundant throughout the last mile of the canyon at this time of year. They have been observed by the hundreds at nearly every level in the canopy on some days. Rose–breasted Grosbeaks are often seen in large numbers during migration. Other species to look for at this time include Kentucky, Cerulean, Nashville, Black–throated Blue, Black–throated Green, and Canada warbler, Hermit Thrush, Louisiana and Northern waterthrush, Winter Wren and the Redstart.

Winter Wrens are quite common during the breeding season, throughout the first three miles of the canyon. If you stay overnight in any of cabins 1–13, you may be greeted by the beautiful song of the Winter Wren, early in the morning.

Several species of warblers reside in the canyon during the nesting season, June–July. Black–throated Blue, Black–throated Green, Kentucky, Louisiana Waterthrush, Parula, Ovenbird and Hooded warblers are abundant. The Yellow–throated and Swainson's warblers are less common. Redstarts are abundant where fallen trees have created openings in the canopy. The Common Yellowthroat, several species of woodpeckers, Ruffed Grouse, Wild Turkey and Broad–winged Hawks are relatively common.

In winter, the numbers of birds are low. Due to the large timber, the canyon is a good place to observe Brown Creepers. Other birds to look for include Golden–crowned Kinglets, White–breasted and Red–breasted nuthatches, Tufted Titmice, Dark–eyed Juncos, Carolina and Black–capped chickadees, Downy and Pileated woodpeckers, Yellow–shafted Flickers, flocks of Wild Turkeys, Ruffed Grouse, and Barred Owls. One might also see flocks of Yellow–rumped Warblers and an occasional Ruby–crowned Kinglet.

In 1966, an 18 acre lake was constructed on a tributary of Glade Creek. Within the last thirty years, a rich assortment of aquatic plant life has accumulated around the edges of this lake, providing cover and food for numerous frogs, salamander, snakes, mammals, and birds. Spotted and Solitary sandpipers, Pied–billed Grebes and Great Blue Herons are common in spring and summer. A pair of Woodcock have been observed in the woods at the upper reaches of the lake. Other summer birds are Black–throated Blue Warbler, Gray Catbird, Eastern Towhee, Belted Kingfisher, Great–crested Flycatcher, Scarlet Tanager, Hooded Warbler, Solitary Vireo, Yellow–billed Cuckoo, Eastern Wood–Pewee, Eastern Phoebe, Rough–winged Swallow, and Barred and Eastern Screech–owl. Scaup, Pintail, Mallard and Black duck, Hooded and Common merganser, loons and other aquatic birds may be see in fall to spring.

Beavers are common throughout the region. Numerous old beaver dams exist along the more gentle gradient of Glade Creek, just east of the park. An old dirt road continues from the grist mill, upstream along the creek. There beaver pond areas are good for wading birds like Greater Yellow–legs, sandpipers, Red–shouldered and Cooper's hawks and Wood Ducks. Tree Swallows have also been noted here.

From upland trails, with spectacular views of the canyon, like Skyline, Island in the Sky and Manns Creek Canyon, one is likely to see Common Ravens, Broad–winged and Red–tailed hawks, and American Crows soaring across the chasm, or spiraling upwards on rising warm air currents. Yellow–throated Warblers, Solitary Vireos, Carolina Chickadees, Downy and Hairy woodpeckers, White–breasted Nuthatches and Hooded Warblers are commonly seen in the pines, oaks, and sassafras trees clinging to the rocky outcrops.

Flocks of Yellow–bellied Sapsuckers, Yellow–rumped Warblers, Cedar Waxwings and kinglets fill these trees during fall migration. During the winter the numbers of birds and species drops.

The park offers cabin rentals and camping, April – October. The original cabins have huge stone fireplaces in them though they are equipped with modern kitchens and baths.

Kanawha Falls Area

Hullet Good

Located in Fayette County along US 60, a water impoundment at **Glen Ferris** extends upstream for two miles to **Gauley Bridge**, where the New River and the Gauley River converge to create the Kanawha River. Often, impressive waterfowl flights occur in season, and a variety of ducks will be found wintering. Common Golden–eyes, Buffleheads, Hooded Mergansers, Mallards, and Black Ducks winter with regularity. A variety of species in large numbers will often be found during adverse weather in winter and spring.

On November 24, 1995, 7,500 waterbirds were observed

on the Kanawha River between Charleston and the Gauley Bridge in Fayette County. This influx was associated with adverse weather conditions during the passage of a low pressure system across the state the day before. About 5,000 were seen at Kanawha Falls. The following were seen in small numbers: 6 Common Loons, 40 Pied–billed Grebes, 6 Double–crested Cormorants, 12 Tundra Swans, 1 American Black Duck, 25 Northern Pintails, 22 Green–winged Teals, 10 Northern Shovelers, 2 Wood Ducks, 50 Canvasbacks, 11 Oldquaws, 5 Surf Scoters, and 15 Bonaparte's Gulls. In greater numbers, 300 Horned Grebes, 100 American Wigeons, 500 Redheads, 60 Ring–necked Ducks, 2,800 Lesser Scaups, 1,000 Buffleheads, 250 Ruddy Ducks, 500 Hooded Mergansers, 900 Red–breasted Mergansers, 300 American Coots, and 400 Ring–billed Gulls were seen.

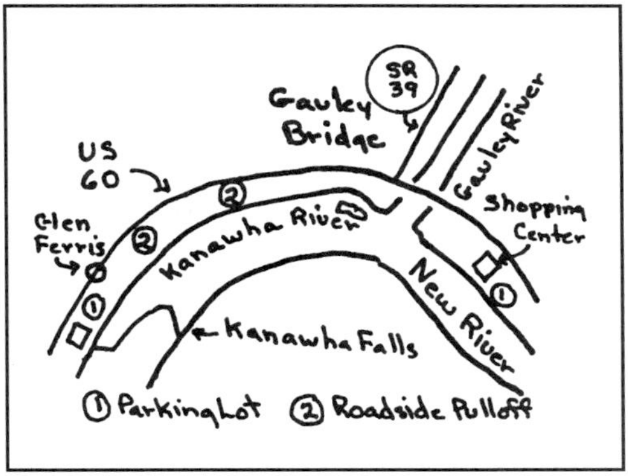

Bluestone State Park
Bluestone Lake–Dam and New River
Ann & Grady McRae

To reach this area from I–64 (Hinton, WV) take the Sandstone exit 139 following SR 20 (10 miles) into Hinton, (see map p-73) then continue on SR 20 south to the entrance of Bluestone State Park (CR 20/2). Or coming from the south on I–77, take the Athens exit and continue north on SR 20 until you reach the entrance area of **Bluestone State Park**

(21.1 miles).

Migrating waterfowl may be seen from November through April. Pied–billed Grebe, Great Blue Heron, Canada Goose, Wood Duck, Mallard, Black Duck, Common Goldeneye, Bufflehead and Hooded Merganser are quite common, and half a dozen more species of ducks, other species of grebes, mergansers, gulls, some shorebirds, and Double–crested Cormorants are possible. Osprey, and Bald and Golden eagles have been seen in this area's Audubon Christmas and spring bird counts (Pipestem Christmas Count–Bibbee Nature Club).

To do an auto tour from the mouth of the **Bluestone River down the New River to Sandstone Falls** (approximately 11.4 miles), start at the entrance of Bluestone State Park along the dam side of SR 20 south. A scope is best to scan the waters below in the lake. Follow the road along the edge of the lake (going north) till you come to a junction (at 3.2 miles) and turn right over the bridge over New River. At the end of the bridge is another junction. Turn right and go 500 feet and turn right on Maple Street (at the Exxon Station), go right again at the next street, and follow the lane around Riverside Drive to Bellepoint Park below the dam. Scan the rapids and shallow waters for foraging and fishing waterfowl.

Return to SR 20 across the same bridge. Continue down river, stopping at frequent intervals. (The Dairy Queen's dining room windows offer an excellent view of the river where hungry birders can chow down and work on their bird list simultaneously). At the next SR 20 Bridge (do not cross) turn down Madam's Creek (CR 3/21), crossing a small bridge and turn right to follow New River to Sandstone Falls, stopping at the many pullovers.

Sandstone Falls, a unique area – Check with the National Park Service in Hinton for additional information. Several rare plants (Kentucky coffeebean tree, purple cliffbreak, and toothache tree) have been identified here. A boardwalk below the falls provides access to a nature trail. In winter, Winter Wren, Hermit Thrush, and Cedar Waxwing and in summer vireos, warblers and other songbirds can be seen.

Bluestone State Park – In spring and summer the road into the park and campgrounds, and the trail up into the

Bluestone Public Hunting and Fishing Area abound with bird song testing the eyes, ears, and stamina of the most dedicated birder. A walk around the RV campground and picnic areas is usually most productive. Northern and Orchard orioles, Rose–breasted Grosbeak, Scarlet Tanager, Indigo Bunting, Wild Turkey, Ruffed Grouse, several species of woodpecker and flycatcher can be seen. At night listen for the Screech, Barred and Great Horned owls.

Bluestone National Scenic River

Jim Phillips

The Bluestone River Canyon is located within the **Bluestone Wildlife Management Area** between Pipestem and Bluestone State Park off SR 20.

The best way to enter the Bluestone Canyon is to ride the aerial tram in **Pipestem State Park.** The tram descends about 1000 feet from the Canyon Rim Center to the Mountain Creek Lodge. The tram is in operation from May 1 through October 31. Check at the main lodge for specific operating times and current fees.

It is an eight mile hike to **Bluestone State Park** along the scenic Bluestone River from the Mountain Creek Lodge at the base of the tram. Begin by following the **River Trail** from the Lodge down stream. The first section of the trail takes you through deciduous forest along the river. At about 1.5 miles, the trail enters an open area. The vegetation here consists of apple trees left from a farm lot, dogwood

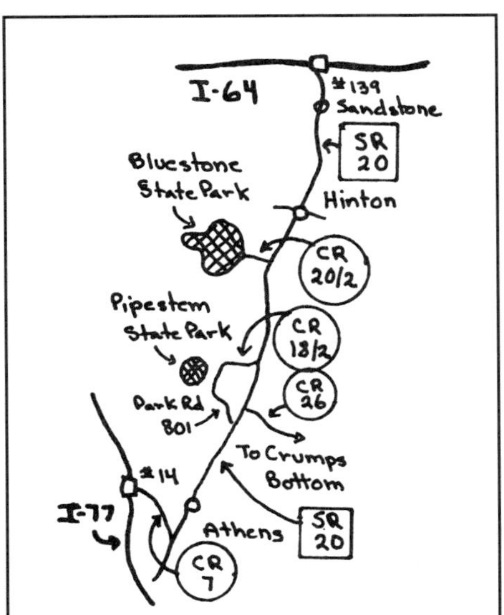

trees, and blackberries and raspberries. At the other end at this old farm lot, the River Trail takes a sharp left turn up Pilot Ridge and leaves the river. In order to continue along the river you should walk straight ahead on an unmarked old roadbed. This road is still used by officials of the Bluestone Wildlife Management Area. This trail is in a public hunting area.

When you are presented with a choice of trails to follow, always pick the trail closest to the river and follow the stream to Bluestone State Park. About 5 miles from the Mountain Creek Lodge, you will have to cross the Little Bluestone River. In Bluestone, it is about three miles to the gate marking the boundary, is a good place to park a shuttle vehicle. From this point you are about 15 miles by road from Pipestem State Park.

Birds you may see along the way include Belted Kingfishers, Common Ravens and Red–tailed Hawks. Ospreys and Golden Eagles are also possibilities. You can see Great Blue and Green herons, Wood Ducks, American Woodcock and Spotted Sandpipers in the summer. Common songbirds along the river include Louisiana Waterthrushes. Wild Turkey and Ruffed Grouse are common game birds in the area.

Bluestone Wildlife Management Area

East of the New River from Bluestone Dam to Virginia Border

Doug Wood

Bluestone Wildlife Management Area, located in Summers and Monroe counties, features **New River Valley** as its primary attraction. Areas accessible to vehicles are maintained by the WV Department of Natural Resources as cropped fields or grass–covered fields. Areas along the river valley between these more developed areas are maintained as remote habitat. For the birder who wishes to venture off the road, several unmarked and unsigned hunter and angler trails wander through the remote areas, mostly very near the river. All roads located near the river may be slick during wet weather. Often in winter and spring New River floods the lower portions of these roads, leaving layers of sand and silt which are not easily negotiated in anything other than 4WD vehicle.

From Beckley, take I–64 east to Sandstone/Hinton exit 139. Take SR 20 south to Hinton. From Hinton, drive east on SR 3 to Junction with SR 12, south towards Peterstown. Pass junction of SR 12 and SR 122, remaining on SR 12 for another 1.5 miles to a sign on the right "Bluestone Wildlife Management Area", and a road sign "Indian Mills Road CR 21/2". Turn right onto Indian Mills Rd., which is a paved 1 1/2 lane road, and follow for 2.8 miles to **Area Headquarters**. There is a camping area here in a large bend in Indian Creek. Parking is ample and the habitat is a field/shrub/park mix along with riparian forest along Indian Creek. Chestnut–sided, Yellowthroat, Blue–winged, Kentucky and Yellow warblers can be seen in spring and early summer. Typical scrub species such as Indigo Bunting, Song Sparrow, Cardinal, Brown Thrasher and Towhee are common. Yellow–billed Cuckoo can be heard here. Keep your ears tuned for Golden–winged Warbler which is heard and seen there occasionally.

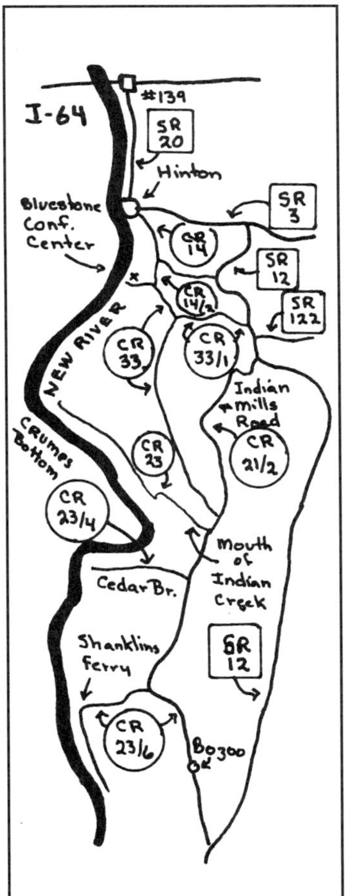

Along Indian Creek look for Green and Great Blue heron, Louisiana Waterthrush and Belted Kingfisher. Red–tailed and Red–shouldered hawks, Turkey and Black vultures are commonly seen, while Great

Horned Owls can be heard at night.

Continue on paved road from HQ area for 2.5 miles to road on right and the sign "Mouth of Indian Creek Camping". Turn right and follow gravel road 0.4 miles to a fork, remain right and follow along fields and, eventually, the New River for some excellent opportunities to see migratory waterfowl in season. The road can be slick and muddy during extended wet periods. *High centered vehicles with snow treads are recommended for wet weather travel.* Good opportunities to see field and scrub species along the field edges. Kingbird and Cedar Waxwings are abundant. To the right of the road are dry, oak–hickory, and pine woodlands with Scarlet Tanagers, Ovenbirds, Worm–eating and Cerulean warblers, Wood Thrushes, Red–bellied and Pileated woodpeckers. Cooper's Hawks are often seen hunting along the woodland edges. Greater Scaup, Hooded Merganser, Wood Duck, Blue–winged Teal, Wigeon and Bufflehead are seen on the river during migration. Backtrack to paved road and continue south for 1.9 miles remaining right at intersection in 0.4 miles and passing gravel road to Cedar Branch Camping Area in 0.7 miles to sign and road on right "Shanklins Ferry Camping Area". While traveling between **Shanklins Ferry Road** and **Cedar Branch Road** on paved road, watch in fields along fence rows for Grasshopper and Savannah sparrows which occasionally nest in the pastoral setting. Barn Swallow is a common species as are Kestrel, Meadowlark and Mourning Dove. Turn right onto gravel road at "Shanklins Ferry Camping Area" sign and follow to fork. Take left fork which eventually leads to New River beside a series of islands. Excellent opportunities for viewing waterfowl previously mentioned. Wood Ducks nest there as do Canada Geese. Black Ducks can be seen even in summer. Ospreys are seen flying between the deeper pools of the New River where they prefer to do their fishing. Kingbirds and Cedar Waxwings take advantage of the abundant aquatic insect hatches as do Tree Swallows and Great Crested Flycatchers. The proximity of maturer oak trees to New River make this area an excellent habitat for Wood Ducks. Mature riparian habitat attracts Yellow–throated Warblers. Wooded cliffs on both sides of the valley make this area a good spot for seeing Ravens and both vultures. Often Black and Turkey vultures will soar together in the area. It is probable that vultures and ravens nest in the area.

The road into Shanklins Ferry ends along the New River at a locked gate. Backtrack to the paved road and continue south through pastoral land in Monroe County. On the left at 0.1 mile is Bozoo store at the crossroads known as Bozoo. The country store is well stocked with snacks and staples for hearty meals on the road while birding. Just beyond the store are several road intersections. The main road CR 24 is marked and is a two lane road. Follow it toward Peterstown for 4.8 miles to SR 12. Along this road watch for field species, keeping in mind that occasionally Shrike, Grasshopper and Savannah sparrows are seen in the area. At the intersection with SR 12, turn left and head north back into Hinton.

Route from Princeton, WV. and Blacksburg, Va. From Princeton follow US 460 east to Rich Creek, Va. and US 219. From Blacksburg go west on US 460 to same location. Follow US 219 north to Peterstown W.V. and junction with SR 12. Follow SR 12 north for 0.8 miles to Bozoo Road. Turn left onto Bozoo Road and follow for 4.8 miles to crossroads and country store at Bozoo. Turn left onto Shanklins Ferry Road (CR 24/3) and follow for 0.1 mile to sign and gravel road on left "Bluestone WMA Shanklins Ferry Camping Area". Follow directions given above in reverse order for the remainder of the trip. When coming back onto SR 12 at the intersection with CR 21/2 Indian Mills Road, turn right on SR 12 and head back towards Peterstown.

Bluestone Conference Center

Jim Phillips

The Bluestone Conference Center is a Presbyterian Church Camp located on the east side of The New River about 7 miles from Forest Hill, West Virginia, on SR 12. To reach the center (see Bluestone WMA map on page 75) from Bluestone Dam, go 0.2 mile north on SR 20 and then take SR 3, turn right onto Cedar Avenue. Follow Cedar Avenue 0.1 mile and turn left onto Zion Mountain Road (CR 14). This is a narrow, winding road which offers a "bird's–eye" view of Bluestone Dam and the town of Hinton.

After traveling on CR 14 for 3.5 miles, you will come to

an intersection where CR 14 becomes Harvey Road and goes to the left. You need to turn right and follow the Low Gap Church Road (CR 14/2). After 2.8 miles, you will come to a rolling pasture which sometimes has Horned Larks in it. Also look here for Red–shouldered Hawks and Wild Turkeys.

From this pasture, drive 2.4 miles and turn right at the church. This paved road passes through the Presbyterian Church Camp and then becomes dirt and gravel as it heads to a 3–way intersection. The road straight ahead leads to the river, the jeep trail to the right is not recommended for passenger cars; the road to the left goes through the riverside campground, and ends after 0.8 mile.

In this vicinity look for Eastern Kingbirds, Red–bellied Woodpeckers, and Tree Swallows in the summer. Near the water look for Wood Ducks, Green Herons, and Canada Geese. In the spring, you may see as many as two dozen species of wood warblers here. During the fall, winter, and spring, your waterfowl list may include Common Loon, Pied–billed and Horned grebes, American Coot, and several species of ducks. gulls and sandpipers also occur here.

Back at the intersection near the church, you can return to Hinton the way you came or you can turn right and return via the town of Forest Hill. If you head to Forest Hill, you will reach an intersection after 1.7 miles. A left turn takes you 4.2 miles to Forest Hill and SR 12. Check the fields along the way for Loggerhead Shrikes and American Kestrels. Check the farm ponds in the spring for shorebirds. Hinton is to the left on SR 12; Peterstown is to the right.

At the intersection 1.7 miles from the church, a right turn will take you toward Indian Creek and Shanklin's Ferry.

Brush Creek Watershed

Jim Phillips

From the Mercer County Court House in Princeton, go south on SR 20 for 4 1/2 miles. **Glenwood Recreation Area** is on the left side of the road. The park is usually in operation from Memorial Day to Labor Day and the admission price is 50 cents for adults and 35 cents for children. Depending on the weather, the park may open earlier or remain open later. During the off season it is permissible to park at the gate and walk in. Many local residents use the area for walking their dogs and jogging.

The park occupies a little over 300 acres with a 55–acre lake. The park road is about two miles in length and offers access to the water which consists of two pools connected by a narrow channel. Besides the open water habitat, the park provides alders, cattails, wood lots and open fields. The best time for migrant waterfowl is March, April, October and November. Unusual birds that have been seen here include Double–crested Cormorants, Willets, Black and Common terns. The wooded areas are especially good for wood warblers and Vireos in late April and early May. The best time for unusual migrants is during and after severe storms. Under these conditions as many as 12–14 species of waterfowl may appear. Summer birds include Green Herons, Wood Ducks, Acadian Flycatchers, Barn Swallows, Ovenbirds and Common Yellowthroat. In winter look for Great Horned Owls, Wild Turkeys and Dark–eyed Juncos.

To continue the tour of the **Brush Creek Watershed**, return to the park entrance. Turn north on SR 20. At 0.9 mile

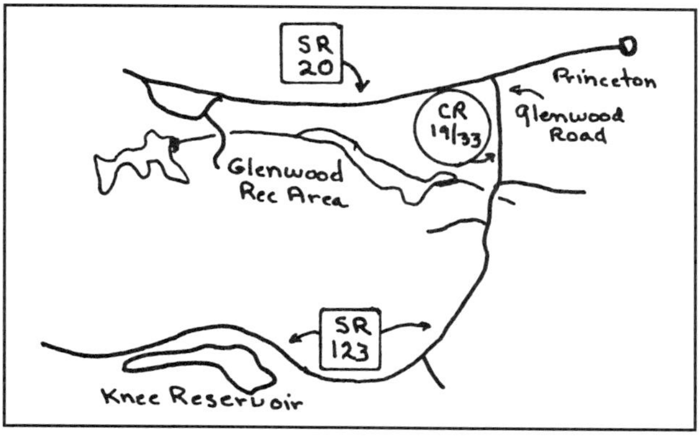

turn right on the Glenwood Road. Drive 1.7 miles to a three–way stop. A left turn takes you to US 19 and 460. A right turn takes you 0.2 mile to the Mercer County 4–H Camp and the Methodist Home for the Aging. Near the camp parking area you may view the dam end of the Glenwood Park Lake, however, park officials prefer that you not enter the property from here when the park is open. This is a handy stop when you do not have time to walk around the lake.

Return to the three–way stop sign and turn right. Go 0.1 mile and turn left. Two–tenths of a mile down this road you will find a wide spot to pull over in front of a sewage lagoon. The lagoon has been divided into two ponds. Black Terns, Common Snipe, Spotted and Solitary sandpipers and both yellowlegs may be found during the spring. Red–winged Blackbirds, Yellow Warblers, Common Yellowthroats and Willow Flycatchers nest here. In August and September look for as many as 100 Wood Ducks. In winter look for Gadwall, Wood Ducks, American Coots and American Widgeon.

To continue the Brush Creek Watershed tour backtrack the 0.2 mile and turn left. After about 1 1/2 miles this road becomes Route 123 west. Two and one tenth mile ahead you will come to the **Kee Reservoir**. Most of the water is on the left side of the road. The lake is about 70 acres in size and is the largest in the watershed. Another 0.3 mile leads you to the parking area. Common Loons, Double–crested Cormorants, Horned Grebes, Snow Geese and Canvasbacks have been seen, here with some regularity. A spotting scope will be helpful. Gulls, terns and Osprey are seen in migration. Look for Green Herons, Wood Ducks and Broad–winged Hawks in summer.

Other lakes exist in the Brush Creek Watershed but this tour allows you to visit the more accessible areas in a relatively short period of time.

Crump's Bottom/Bull Falls Area of Bluestone Wildlife Management Area

Jim Phillips

Crump's Bottom is located midway between Princeton, Mercer County and Hinton, Summers County on the west bank of Bluestone Lake (New River) and east of SR 20. To reach this area, (see Bluestone SP Map), from I–64 take the Sandstone Exit and follow SR 20 south toward Pipestem State Park (about 25 miles). To reach this area from I–77 take the Athens Exit and follow the signs to Pipestem State Park (about 13 miles).

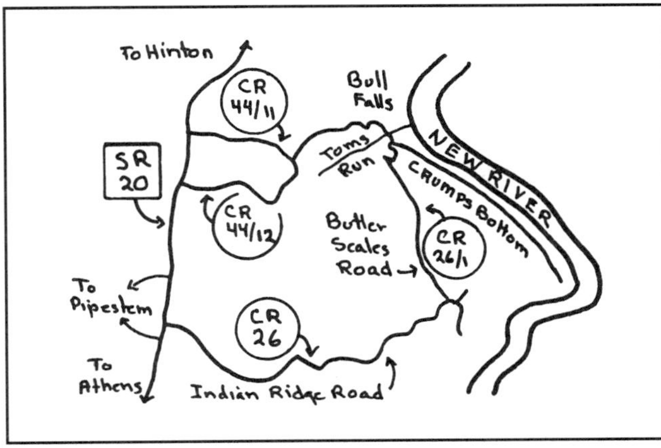

Crump's Bottom offers some of the best birding in the New River area. From late April to mid–May, you can list two dozen species of warblers, five species of vireos and six species of swallows. As many as 17 species of warblers occur here in the summer including Yellow–throated, Prothonotary and Cerulean. Also in the summer look for Green and Great Blue heron, Black–crowned Night–herons, Ospreys, Northern and Orchard orioles. Common Goldeneyes, Buffleheads and Hooded Mergansers are common wintering waterfowl. Bald Eagles and Rough–legged Hawks have also been seen in winter. Tundra Swans, Common Loons and Double–crested Cormorants occasionally add spice to the bird list.

From the entrance to Pipestem State Park on SR 20, go 0.2 miles north and turn right onto Indian Ridge Road, (CR 26), and follow it 5.3 miles, then turn left on the Butler Scales Road, (CR 26/1). This road will follow Tom's Run to Crump's Bottom about 2.4 miles. The first .25 mile of the road offers large hilly fields that are food for American Kestrels, Eastern Bluebirds and Eastern Kingbirds. In season, the remaining section of the road can be good for Acadian Flycatchers, Scarlet Tanagers, Worm–eating Warblers, Barred Owls, Ruffed Grouse and Wild Turkey. During times of heavy rain, snow, or freezing weather, this road my be treacherous. A 4–wheel drive vehicle may be in order! Park near the intersection and walk along Tom's Run. In summer this is a good area for American Redstarts, Louisiana Waterthrushes, Northern Parula, Yellow–throated and Cerulean warblers.

If you turn right at the intersection, you can drive for five miles through Crump's Bottom. You will have to turn around and come back the same way. If mud makes the road impassable you may be better off parking and walking the road or some of the side trails. The first few trails to the left lead to some backwater, oxbow type areas of Bluestone Lake. Check here for nesting Prothonotary Warblers and Eastern Screech–Owls.

At 1.2 miles you will begin to see corn fields on the left side of the road. In the fall before the corn is harvested, scores of migrating Palm Warblers glean insects from the stalks. Look for Canada Geese, Mourning Doves and Wild Turkeys after the corn is harvested.

At 2.7 miles, you will find a small pond located between two corn fields. Look for Spotted and Solitary sandpipers and both species of yellowlegs during their migration seasons. In summer look for American Woodcock, Sora and Virginia rails.

At 3.6 miles, you will enter an area that was burned a few years ago. Look for Chestnut–sided, Golden–winged and Blue–winged warblers and Yellow–breasted Chats.

At 4.8 miles, you reach a gate where you will have to turn around and return the way you came. The area beyond the gate is still within the Bluestone WMA, so you may wish to explore on foot.

When you return to the intersection at Tom's Run, cross the bridge. The Bull Falls campground is on your right. In the winter, the campground is a great place for woodpeckers, Yellow–rumped Warblers and Cedar Waxwings. In the summer, look for nesting Tree Swallows, Warbling Vireos and Eastern Kingbirds.

As you leave the campground, turn right and go up the hill 1.8 miles to a hardtop road. Turn left and drive 0.5 mile to an intersection. Either road here will lead you to SR 20 just north of the Pipestem State Park entrance. The road to the right is Gravely Point Road (CR 44/11). The road to the left is Broadway Road (CR 44/12). Along either of these you are likely to see American Kestrels or Eastern Meadowlarks. Both Gravely Point and Broadway Roads are marked on SR 20 as access to Bull Falls. If the bottom is under water or the roads are really terrible you will encounter locked gates or "road closed" signs. Always use caution in wet weather.

Peters Mountain/Hanging Rock Raptor Migration Observatory

Cindy Ellis

The best Hawk watching in West Virginia may be from **Hanging Rock** on **Peters Mountain** in Monroe County. Where else can one look *down* on a Bald Eagle as it glides through a beautiful valley?

To reach Peters Mountain from Beckley, travel I–64 east to exit 169 at Lewisburg. Drive south on US 219 to Union and turn left on SR 3 to Gap Mills. From the east drive on I–64 west to the West Virginia and Virginia state line and get off at exit 183 and proceed south on SR 311 to Crows and on to Sweet Springs. Then take SR 3 to Gap Mills. Alternately from Bluefield, take US 460 east and exit at Peterstown, WV and take US 219 north to Union, making a right on SR 3 to Gap Mills. At Gap Mills take CR 15 southwest and go 3 miles. A gravel road turns left, up the mountain and may be labeled To Waiteville. Drive 1.6 miles to a sharp bend in the road and park. Walk up a moderately steep 1 mile trail, bearing left at the one fork in the trail (right goes to the Allegheny Trail).

Hanging Rock itself is a jagged outcrop on the ridge of Peters Mountain. Wide valleys north and south provide large areas for the build–up of thermals. This area is along a traditional migratory flight path for hawks from the north traveling across the barrier of the Allegheny Front.

Through 1995, hawk watchers used a renovated tower formerly used by the forest service for spotting fires. Unfortunately, in early 1996, vandals burned the tower. Government officials, especially those connected with the Jefferson National Forest, have rebuild the structure with many improvements.

Traditionally, the hawk count is conducted for the 10 days or two weeks before and after September 15, from mid–morning to late afternoon. For more information contact count census person Jerry Davis in Lewisburg (1–800–983–6374 or 304–647–7000), or in the Huntington area, Wendell Argabrite (304–736–5747), or in the Charleston area, Cindy Ellis (304–586–4135).

Moncove Lake State Park (304–772–3450) provides good camping nearby. Lodging is available in Lewisburg or at a B & B in Sinks Grove (Morgan's 304–772–3638). Mountain View, a very good restaurant is east of Gap Mills; open evenings only.

Often the weather is cool and windy on the ridge, despite warmer conditions in the valley; dress accordingly.

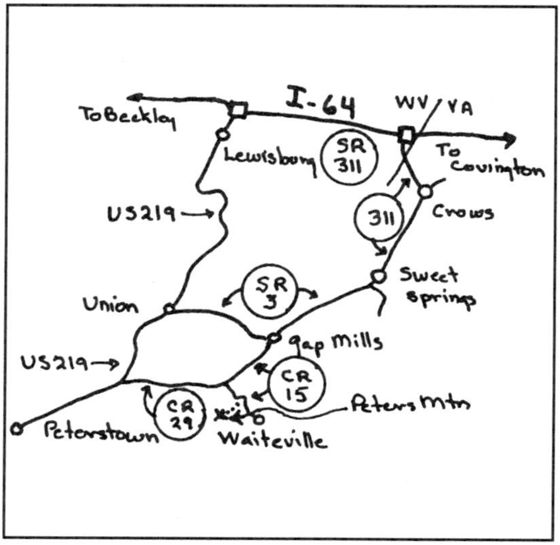

R.D.Bailey Lake Wildlife Management Area

Ronald A. Canterbury

The **R. D. Bailey Lake WMA** is a 19,000 acre public hunting and fishing area located along the Guyandotte River at Justice, Mingo County, West Virginia and in the Allegheny Plateau Physiographic Province (Gilbert Quadrangle). Much of the lake project is located in the southwest portion of Wyoming County near Mingo County. The lake project is a man–made reservoir and dam developed by the US Army Corp of Engineers to reduce flood damage along the Guyandotte and Ohio Rivers. The lake improves water quality, fish and wildlife habitat, recreation, and flood control. The dam is 310 feet high, the second highest in WV. Outdoor recreation at the lake includes boating year-round, a marina, fishing pier, picnic tables and grills, shelters, playground, camping areas, and restrooms.

R. D. Bailey Lake is 63 car miles southwest of Beckley, WV. You can easily access the lake by taking exit 42 off I–77/

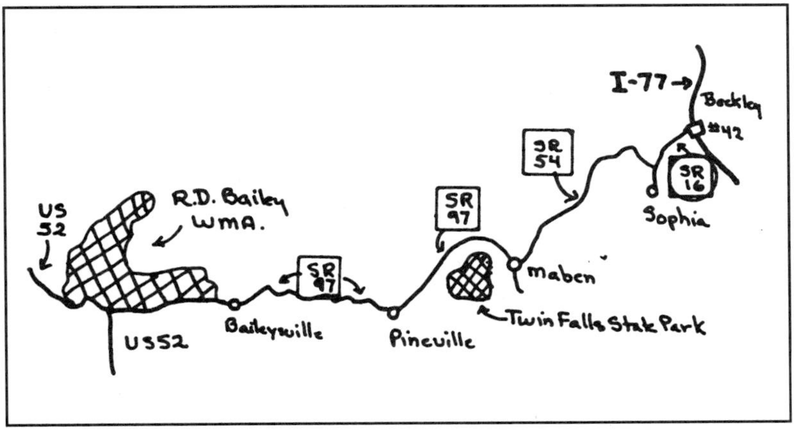

I–64 at Mabscott. Take SR 16 south toward Sophia. Turn right on SR 54 and travel 14 miles to Maben. At Maben, make a sharp right turn on SR 97 and proceed through Pineville and Baileysville. Continue to the intersection of SR 97 and US 52. Turn right on US 52 and go 2.2 miles. Make a right turn and drive up the mountain 1.2 miles to the visitor center where you can pick up brochures and maps about the area.

Some birding localities at R.D. Bailey Lake area include (1) the Bottomland on SR 97 between Baileysville and US 52, (2) Downstream Fishing and Recreation Area, (3) Visitor's Center, (4) Big Branch Day Use Area, (5) Guyandotte Point Boat Ramp and Marina, (6) Long Branch Overlook, and (7) Guyandotte Campground.

(1) **Bottomland** between Baileysville and US 52 – There are numerous birding areas along SR 97. Walking the railroad tracks along SR 97 and the Guyandotte River offers good birding in the bottomlands, old fields, and river edges.

You will hear and/or see Northern (Yellow–shafted)

Flickers in the bottomland. Look for Red–bellied Woodpeckers foraging in the open areas with dead trees. In the forested hillsides, you will find Pileated, Hairy, and Downy woodpeckers. Eastern Kingbirds, Great Crested Flycatchers, Eastern Bluebirds, and Chipping Sparrows are prevalent along the power line right–of–ways. Carolina Wrens, Common Yellowthroats, Northern Cardinals, Song and Field sparrows are numerous in the old fields and bottomlands. Chimney Swifts and swallows are abundant during spring migration in mid–April along the Guyandotte River. Look for Tree and Barn swallows. You will also notice flocks of Cliff Swallows and Purple Martins foraging over the river. Birding at night will disclose how numerous Eastern Screech-owls and Barred Owls are in WV. In late May, you should still be able to hear Whip–poor–wills, a species that has drastically declined from much of its former range.

(2) **Downstream Fishing and Recreation Area** – Leave the visitor center and go 1.1 miles to US 52 and turn right. Go 0.1 miles and make a right turn and then drive 0.2 miles and turn left across the bridge.

Trails along the lake provide access for birding the shoreline and floodplain. In spring look for some late winter visitants such as Yellow–bellied Sapsuckers, Hermit Thrushes, and Winter Wrens. Noteworthy will be mixed flocks of neotropical migrant songbirds, such as Least Flycatcher, Yellow–throated and Warbling vireos, Yellow–rumped and Blackburnian warblers. In late May look for Bay–breasted and Black–throated Blue warblers. (Please take note of any Olive–sided Flycatchers during the summer months and report your sightings to the WVDNR in Elkins.) Eastern Towhees are abundant. Listen for edge species such as Field and Song sparrows, Indigo Buntings, Common Yellowthroats and a few Blue–winged Warblers. Others include Gray Catbirds, Yellow Warblers and White–eyed Vireos. In fall, migration brings large numbers of Magnolia, Cape May, Tennessee, Black–throated Green, Black–throated Blue and western Palm warblers.

(3) **Visitor's Center** – The Visitor's Center offers an overlook of the dam and lake as well as a view of the upland mixed deciduous forest surrounding the lake. The upland forest is dominated by oaks, maples, and hickories. Hike the trails and note foraging Blue–gray Gnatcatchers during

spring migration. Also look for mixed flocks of spring migrants which may include Gray–cheeked and Swainson's thrushes, and Blackpoll and Cape May warblers. One particular flock observed on 9 May 1996 had 4 Blue–gray Gnatcatchers, 1 Yellow–throated and 6 Red–eyed Vireos, 5 Tennessee, 8 Blackpoll, and 2 Cape May warblers.

(4) **Big Branch Day Use Area** – Return to US 52 and turn left. Proceed 2.2 miles to SR 97 and turn left again. Continue on SR 97 for 2.0 miles to Coal Mountain Road (CR 6/2). Proceed 0.1 miles on CR 6/2 and turn left to reach Big Branch Day Use Area.

During breeding season, look for Carolina Chickadees nesting in abandoned woodpecker holes. American Redstarts and Wood Thrushes, Downy Woodpeckers and Great Crested Flycatchers are also abundant.

(5) **Guyandotte Point Boat Ramp and Marina** – Follow the Big Branch Day Use Area directions. At 0.8 miles from SR 97 on CR 6/2, turn left to the boat ramp.

This is a good place to use a spotting scope to search for waterfowl during the fall and winter. In late October through mid–November, look for ducks such as Green–winged Teal and Redhead. You may see Hooded Mergansers and American Coots. Mid–November through mid–February, look for Ruddy and American Black ducks.

(6) **Long Branch Overlook** – Again follow the directions to Big Branch Day Use Area. At 1.0 miles from SR 97 on CR 6/2, you will come to the scenic overlook on the left. The drive along Coal Mountain Road (CR 6/2) to the scenic overlook will allow you to view the rugged mountain slopes of WV and to note the extreme elevation differences between the floodplain and knobs. The elevation at R.D. Bailey Lake ranges from 960 feet at the dam to 2,010 feet at some of the peaks.

Birding is generally very productive along Coal Mountain Road. At the large rock outcrops look for nesting Eastern Phoebes after March. A few Dark–eyed Juncos will linger until late April or early May and you will hear numerous White–throated Sparrows singing in mid–April in the old fields.

(7) **Guyandotte Campground** – To find Guyandotte Campground, leave the visitor center and return to US 52 and turn left. Go 2.2 miles to SR 97. Turn left on SR 97 and

go 5.8 miles and make a left to the campgrounds. In the spring and fall look for migrant warblers and vireos. In spring listen for the singing Red–eyed Vireos, Scarlet Tanagers, Hooded Warblers, and Least Flycatchers. In winter look for foraging Ruby–crowned and Golden–crowned kinglets, White–breasted Nuthatches, Carolina Chickadees, and Tufted Titmice in mixed flocks.

In summer, you will find both Great Blue and Green herons. Look for Mallards and Wood Ducks, Golden–winged Warblers and Common Yellowthroats in the marshy wetlands, and Acadian Flycatchers on many of the tributaries. Sloped hillsides offer Black–and–white and Worm–eating warblers. Brushy second growth areas offer Willow Flycatchers and White–eyed Vireos. Also in abundance are Blue Jay, Tufted Titmouse, Song Sparrow, Northern Cardinals, Indigo Buntings, and Eastern Towhees. Listen for Red–bellied Woodpeckers and White–breasted Nuthatches. Northern Rough–winged Swallows are abundant as well as Purple Martins, House Wrens and 6–9 pairs of Prothonotary Warblers nesting in the bottomlands. Some rather uncommon species include Northern Parula, Cerulean, Prairie, Kentucky, and Worm–eating warblers, Yellow–breasted Chats and Louisiana Waterthrushes.

Twin Falls State Park

Ronald A. Canterbury

Twin Falls State Park is a 3,776 acre park located in Wyoming County, West Virginia.To reach the park (see R.D. Bailey WMA map on page 85), take exit 42 (Sophia/Mullens) at Mabscott off I–77/I–64 and travel southwest on SR 16/97 towards the towns of Sophia and Mullens. Go 2.7 miles to the junction of SR 97 and SR 54 and make a right turn. Travel 14 miles on SR 97 to Maben, where SR 97 takes a sharp turn towards Pineville. Continue 6 miles on SR 97 and turn left. You will travel 2 miles to Twin Falls.

Twin Falls is in the Allegheny Plateau Physiographic Province and is characterized by a mixed deciduous forest dominated by oak and hickory. Eastern hemlock is common along the coves and ravines. The highest point in the park is 2,409 feet.

Twin Falls has many recreational opportunities such as hiking, golfing, fishing, swimming, and nature programs. The park has camping with or without electricity, bathhouse with hot water, laundry, and deluxe cottages. The lodge has a restaurant, 20 guest rooms, conference center, gift shop, and nature center. A map can be obtained at the main headquarters of the lodge.

Upon entering Twin Falls, stay on the main road until you reach the signs for the **Lodge**. After traveling 2.7 miles, turn left at the lodge/conference center sign. The mixed deciduous forest and old fields surrounding the lodge and golf course offer some good birding. You will enjoy the usual resident species such as the Red–bellied Woodpecker, Carolina Wren, Song Sparrow, White–breasted Nuthatch, Tufted Titmouse, and Northern Cardinal. In spring, look and listen for migrating Bobolinks along the golf course. You will also find Cedar Waxwings, Gray Catbirds, Wood Thrushes, and many American Robins along the roadsides and bushy woodland edges bordering the golf course and lodge area. In fall, this is a good place to find migrant warblers such as Pine, Chestnut–sided and Black–throated Green.

Other birding locations include the **cottage area, maintenance area**, **White Horse Knob Tower** area, **old fields along the museum, woodland picnic area, the campground area, and the pioneer farm**. To find these areas from the lodge, go back to the main road and turn left. Go 0.2 miles and turn right. The road ends at the pioneer farm. You will find good birding if you take the time to make most of these stops. On August 19, 1995 by covering the above mentioned areas, the more interesting birds seen were as follows: at the lodge, Red–eyed Vireo, Cedar Waxwing, Pileated Woodpecker, Acadian Flycatcher, Black–billed Cuckoo, Black–and–white Warbler, and Downy Woodpecker; at the museum, campground and picnic area, Wild Turkey, Blue–gray Gnatcatcher, Ruby–throated Hummingbird, Field and Chipping sparrows, Indigo Bunting. Along **Falls Trail**, Hooded, Parula, Black–throated Green, and Nashville warblers, and Blue Jay. Along **Twin Oaks Trail**, Blue–winged, Northern Parula, Black–throated Green, Mourning, Prairie, Hooded and Black–and–white warblers, Eastern Towhee, Baltimore

Oriole, and Gray Catbird. Along the **Cemetery Trail**, Eastern Phoebe and Worm–eating Warbler; and at the Pioneer Farm, White–eyed Vireo, Indigo Bunting, Tufted Titmouse, Worm–eating Warbler and Eastern Bluebird.

The park offers excellent woodlands for forest–interior species, old fields and brushy edges for the edge species, and grassland birds along the golf course. Each of the nine trails can be easily found with a park map. The **Buck Run Trail** offers steep rugged terrain, rhododendron, and forest species such as Cerulean, and Hooded warblers, Summer and Scarlet tanagers and Ovenbirds. The **Cliffside Trail** has numerous birding areas. Take time to explore **Bearwallow Trail** off the Cliffside Trail and the **Canada and Buzzard Cliffs**. You will find the Carolina Wren, Eastern Phoebe, Golden and Blue–winged warbler and American Redstart in the summer months. The rolling terrain of the **Falls Trail** is a good place to find the Louisiana Waterthrush and Kentucky Warbler. The **Hemlock Trail** leads through rhododendron and mature hemlock where the Ovenbird and Hooded Warbler can be found. The **Huckleberry Trail** is a good place to find numerous species of ferns and wildflowers during the spring. Also listen for migrating Solitary Vireos in late March. This trail passes the **Pioneer Farm** and one should see Blue–gray Gnatcatchers, Red–eyed Vireos and Yellow–rumped Warblers in the spring. Ruffed Grouse are common along all the trails but more frequently seen along the **Nature Trail**.

Where to Look for the Swainson's Warbler

Wendell Argabrite

The starting place, in your hunt for the Swainson's Warbler, is **US 19 in Fayetteville**, West Virginia. Take US 19 north out of Fayetteville and cross the **New River Gorge** Bridge. From the north end of the bridge, go 0.6 miles to Lansing Road on you right. Turn right on Lansing Road. At 0.3 miles Lansing Road narrows and people drive in the middle of the road, so **be careful**. One mile from US 19, there is a parking area on the right. Pull in and park. You can sometimes hear them singing from the parking lot. Take the path to a small creek (about 30 yards). There is usually a pair of Swainson's Warblers nesting in this area.

If you don't see them here, continue on **Lansing Road** for 0.4 miles. Stop at a pull off on your right. Park and take the path down the hill to a fork in the trail (about 50 yards). Take the right fork to the creek (about 20 yards) and play your tape. If you do not have any luck at the right fork, try the left one. Good luck!

The Swainson's Warbler was also seen and heard at **Panther State Forest** in McDowell County, in the New River/Greenbrier Valley Region, in May 1997. From the north end of the park, start looking and listening along the road near the Panther swimming pool. The creek , rhododendrons and evergreens parallel the road and are perfect habitat for this bird that says "see, see, see, sistersville!" very loudly, resembling the Louisiana Waterthrush. Others were heard across the group camp bridge and by walking up the **Crane Branch Trail.**

Good luck again,
Beth Bullard

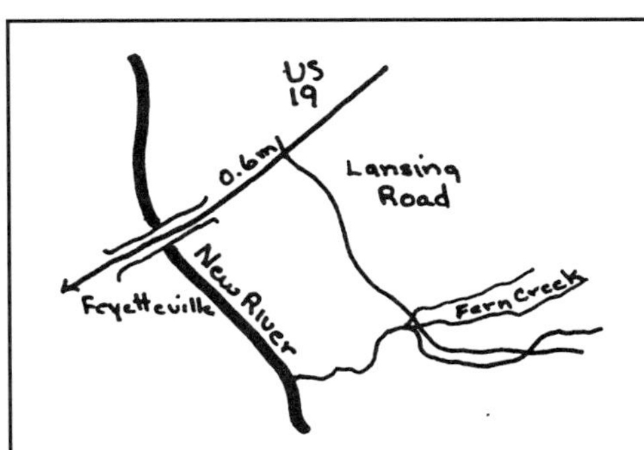

5. The Metro Valley Area

J. Lawrence Smith

The area is essentially the counties of Kanawha, Putnam, Mason, and western Fayette through which the Kanawha River flows, along with the northern portion of Cabell containing the **Greenbottom Wildlife Management Area**.

The valley of the **Kanawha River** is the most prominent natural feature of much of the area. The alluvial bottomland has undergone extensive residential and industrial development on the vicinity of Charleston. The wide bottomland from Nitro to Point Pleasant is used extensively for pasture and crop production. Bottomland along the **Ohio River** is used extensively for agricultural purposes and has undergone nominal industrial development. The area away from the valley is hilly, with the valleys of many of the small rivers and streams being narrow and twisting.

The variation in elevation within the area is from a high of 2,850 feet in southern Kanawha County near the Raleigh County line to 500 feet on the Ohio River in Mason County. The hills and surface topography are the remnants of a peneplain that was originally elevated during late Mesozoic times and has undergone various stages of elevation and erosion. Almost all the exposed rocks are of non–marine origin and consist of sediments laid down on floodplains or in bodies of freshwater throughout much of the Pennsylvanian period. (The Dunkard strata in Putnam and Mason Counties may be of early Permian age). The exposed rocks are mainly sandstones, shales, and thin beds of coal.

The area is within the forest region referred to as "central hardwoods." Forest cover is often extensive along ridges and steep slopes where oak–hickory is the predominant type. Yellow poplar, sugar maple, and basswood are found on deeper soils and in cove situations. Sycamore, silver maple, and sweet gum are often found abundantly along the water courses and in the forested bottomland. Virginia pine, sassafras, persimmon, and black locust are pioneer types on abandoned fields and eroded pastures. Many residences and small farms which have created much brush and edge are located in the hills away from the valley proper.

Weather conditions for the area are rather mild, with the

January mean temperature at Charleston being 38 degrees and the July mean 77 degrees. The average annual precipitation at Charleston is 46 inches. Winter conditions are not rigorous and there is little accumulation of snowfall under normal conditions.

Coonskin Park, Charleston
Hullet Good

Coonskin Park occupies approximately 1000 acres about 2.0 miles northeast of Charleston, WV. Take I–64/I–77 to Greenbrier Street (exit 99). Travel north on SR 114 for 0.9 miles. Turn left onto Coonskin Drive (CR 50/2), at the traffic light. Then follow Coonskin Drive for 0.75 miles to enter the park from the south. Coonskin Park is free and open all year from dawn to dusk. **Warning**: Observe the park speed limit of 25 mph. See page 136 for detailed map.

There are four habitat areas; a shallow **lake** for waterfowl, cattail reed **marsh**, upland hardwood **forest** and an 18–hole **golf course** bounded by the Elk River.

Specialities: Coonskin Park is best in the breeding season. Nesting species include a variety of warblers, such as Kentucky, Ovenbird, Cerulean, Hooded, Yellow–throated and Swainson's, four species of vireos, Belted Kingfisher,

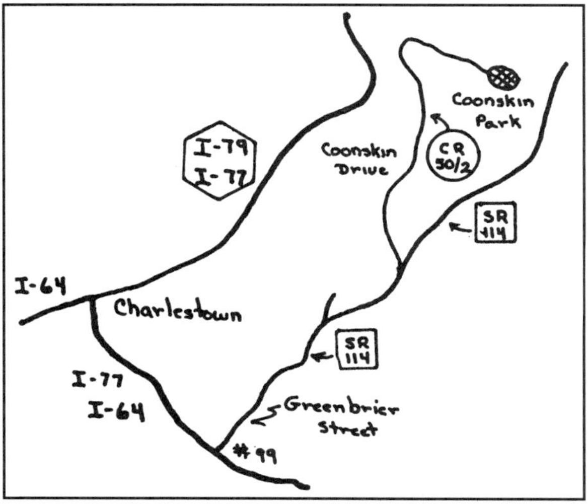

Blue–gray Gnatcatcher, orioles, tanagers, and five species of woodpeckers.

Park Entrance to Herscher Lake (1.0 mile) – While walking slowly along Coonskin Drive, look in the pine and hemlock trees for migratory early winter residents such as Brown Creeper, Hermit Thrush, Sapsucker, Ruby– and Golden–crowned kinglets, Red–breasted Nuthatch, Pine Siskin, Purple Finch and Myrtle Warbler. Permanent residents often seen in the conifers include Carolina Chickadee, White–breasted Nuthatch, Tufted Titmouse, Northern Flicker, and Hairy, Downy, Red–bellied and Pileated woodpeckers.

At the **Gorman picnic area** to your left, look for early spring (March 25 to April 10) migrants: Eastern Bluebird, Robin, Chipping and Field sparrows, Solitary Vireo, Brown Thrasher and Blue–gray Gnatcatcher. In a small ravine just beyond the Gorman area, a nesting Swainson's Warbler has been observed in early May and June for the previous three of four years.

Look for a half acre cattail marsh in front of and to the left of Herscher Lake. The Red–winged Blackbird, Common Grackle, Yellowthroat and Song Sparrow have nested there. American Woodcock and Common Snipe have been seen in the marsh area during the spring and fall migration periods.

At **Herscher Lake** (where there are always many resident Mallards) you can usually see a few Black Ducks and several Canada Geese. Other mid to late winter waterfowl you may see here include Pied–billed Grebe, Ring–necked Duck, Lesser Scaup, Hooded Merganser, American Widgeon, Blue–winged Teal and Coot. In early spring look for Great Blue and Green herons, Belted Kingfisher and Ring–billed Gull. Phoebes, migrant Tree and Rough–winged swallows show up in late March, followed by Eastern Kingbirds, Barn Swallows and Purple Martins in mid to late April. All five woodpeckers (as previously mentioned) may be seen in the trees surrounding Herscher Lake.

Coonskin Nature Trail – After entering the park's main entrance, drive 0.5 miles to Upper Park Road on your right. A scenic nature trail follows Coonskin Branch for 0.75 miles, then crosses the stream and returns to the initial starting point. In early April listen for the singing of the Louisiana Waterthrush and Yellow–throated warblers, which nest here. As you walk up the trail listen for the singing of mid–May nesting warblers such as Parula (in the big hemlocks), Cerulean, Worm–eating, Redstart, Hooded, Ovenbird and Kentucky. The Red–eyed and Yellow–throated vireos are also regular May nesters. Flycatcher species that have nested in the trail area include Phoebe, Wood-Pewee, Acadian and Great Crested. Late winter nesters near the ridge top are Red–shouldered and Red–tailed hawks and Great Horned Owl. Early May and mid September migratory warblers that may be seen along this nature trail are Pine, Tennessee, Magnolia, Cape May, Blackpoll, Blackburnian, Black–throated Green, Bay–breasted and Yellow–rumped.

Exercise Trail to Golf Course (1.0 mile) – The end of the exercise trail begins 0.75 miles from the park's entrance on the right of Coonskin Drive. In late spring look for nesting American Robins, Northern Cardinals, Eastern Towhees, Brown Thrashers, Wood Thrushes, Blue Jays and Chipping Sparrows at the edge (or in) the hardwood forest. In the forest you can expect to see or hear Blue–gray Gnatcatcher, Red–eyed and Yellow–throated vireos, four woodpecker species, and several warbler species, namely; Hooded, Ovenbird, Redstart, Kentucky, and Black–and–white. As the trail winds back down the north side of the hill, Ruffed Grouse may flush from the ravine thickets. While following the narrow field on the valley floor (beside the old CSX railroad tracks) look for Chipping, Field and Vesper sparrows in early April; and Killdeer and Solitary Sandpipers in the wet area in early May. Eastern Towhee, Mockingbird, American Goldfinch and House Finch are often seen in the multilflora rose bushes along the abandoned RR tracks. Early spring/late fall migrant Dark–eyed Juncos, Fox, White–crowned and White–throated sparrows have also been observed in the rose bushes.

Elk River Walk Trail to Coonskin Branch (1.0 mile) – Cross the restaurant and golf course parking lot leading to Elk River. The trail is an old secondary road, paved or gravelled along its way to near the park's boundary line where Coonskin Branch flows into the Elk River. In early spring look for tree nesting Wood Ducks, also seen on Elk River from May through October with their young. One or two Great Blue Herons (non–nesting) can be seen about every month of the year on the Elk. Spotted Sandpiper is common May through August. Arriving in late April/early May to nest in the trees and shrubs are the Warbling, Yellow–throated and Red–eyed and White–eyed vireos. Nesting warblers include Yellowthroat, Redstart, Louisiana Waterthrush and Yellow. In an open area just beyond Herscher Lake listen for the early May singing of the Baltimore and Orchard orioles from the tall sycamore trees. As the trail enters the woods expect to hear or see Scarlet and Summer tanagers. From May through August Cedar Waxwings, Rough–winged and Barn swallows are often seen catching insects along the Elk. Migrant warblers during September include Yellow–rumped, Palm, Nashville, Tennessee, Bay–breasted, Magnolia, Blue–winged and Chestnut–sided. Early arriving winter residents may include Evening Grosbeak, Pine Siskin, Purple Finch, Ruby– and Golden–crowned kinglets.

Eleanor, Buffalo, Winfield, Kanawha River

Cindy Ellis

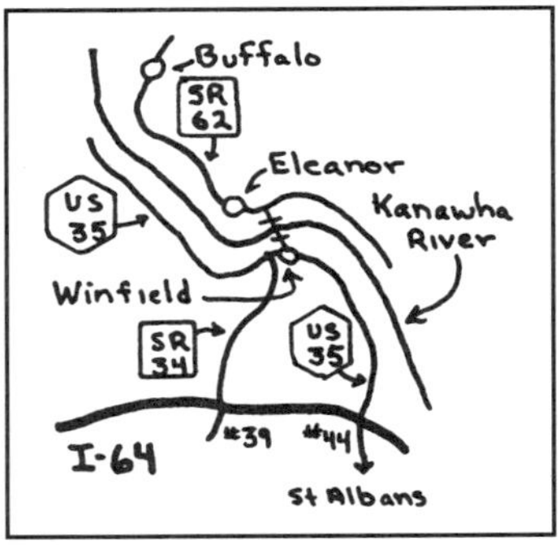

The habitat in the Putnam County Park and along CR 62 and US 35 is largely that of small town, active farmland, and fields returning to forest. A locks and dam at Winfield is undergoing large scale rebuilding and may afford good birding in the future with a large spillway and extensive areas of flat land surrounding. The **Kanawha River** attracts some waterfowl.

Winfield can be reached from exit 39 of I–64 by traveling seven miles northeast on SR 34, or by traveling 10 miles northwest from St. Albans on US 35. Eleanor is one mile from Winfield via the Winfield Bridge across the Kanawha River. Putnam County Park at Eleanor is reached by circling under the bridge and traveling northwest for 1.8 miles. Signs on the left near the county Vocational Center direct one to the park. Buffalo is 10 miles northwest of Eleanor on SR 62.

When stopping along CR 26 and US 35 to bird, use the wider areas of the highway berm.

Putnam County Park – First travel to the park at Eleanor. Near the entrance, to the right of the gate, look for Blue–winged Warblers. The areas left and right just after the gate are good for car birding and seeing local species.

Drive past the historic one–room school building and to the small bridge. This is a good place for Yellow–throated Warblers and orioles. If the first gate on the right beyond the bridge is open, one may drive to the next gate and park near an old picnic shelter. In this area may be found Acadian Flycatcher, Phoebe, Pileated Woodpecker, Wood Thrush, and Ruffed Grouse. Walk along the gravel road to the two gun ranges that are sited so that safety is assured. Range members report seeing Wild Turkey. A pair of Broad–winged Hawks usually nests nearby and one may also see Red–tailed and Red–shouldered hawks, Redstarts, Common Yellowthroat, Carolina Wren, Blue–gray Gnatcatcher, Wood Duck and Belted Kingfisher. There is currently no admittance to the area of locks and dam due to construction.

Other locations – Purple Martins may be seen in the town of **Eleanor**. One established colony has been maintained by a Mr. Perkins on Kopok Street for more than 50 years. Kopok Street is 1.2 miles northwest of the Vocational Center, on the right. On the left in a brushy area adjacent to the railroad, Warbling Vireos have been found.

There is generally good birding north along **SR 62** between Eleanor and Buffalo. Occasionally Horned Larks are seen in the cultivated fields between the highway and the Kanawha River. Blue Grosbeaks have been site faithful for a number of years at one or two spots along SR 62. They have been noted on Breeding Bird Surveys at a railroad crossing 1.0 miles northwest of Buffalo or 2.0 miles northwest of the Mason County line sign on the left. It is usual to note Indigo Buntings near there as well.

Winfield – Seasonal checks of the river near the **Winfield Bridge** may prove productive for migrating waterfowl. Closer views may be gained by traveling the back streets of town; especially Ferry Street, which indeed leads to the **old ferry landing**. Winfield Baptist Church, west of town has a large parking lot and a gazebo near the river's edge.

US 35 is always good for seeing Kestrels. Look there in

winter for Rough–legged Hawks and Northern Harriers. These last two may be noted at any substantial field or windbreak of trees as one travels US 35 for the 30 miles between Winfield and Point Pleasant. In 1994 a large roost of Turkey Vultures stayed there throughout the winter. In 1992 and 1993, Bald Eagles made a brief stay there.

Great Blue Heron Colony– In the spring of 1997, local birders became aware of a heronry of Great Blues. The 30 plus nests are located in two sycamore trees on the south bank of the Kanawha River, but are more visible from the north side. Here are the directions: From the Winfield bridge, travel 12 miles north on SR 62. (Travel through Eleanor and Buffalo; go 1.5 miles north of Buffalo.) On the left is a small white bus shelter. Turn left here on a gravel road leading to a railroad crossing and a small group of houses on the riverbank. Park between the last two residences on the left, a trailer and a white house. Walk between the two properties to the riverbank and look south and upriver. The two trees are on the opposite shore.

Kanawha State Forest
Diane Packett

Comprising nearly 7,000 acres, the **Kanawha State Forest** is located four miles south of Charleston on the headwaters of Davis Creek. To reach Kanawha SF from I–64, exit at US 119 south (Corridor G) at Oakwood Road (1 mile south of the Kanawha River) and follow the well marked Kanawha State Forest signs 6.5 miles to the Park entrance. Nearly all of the area is forested by nearly mature hardwoods. Forest cover ranges from a "cove" type of oak, American beech, and yellow poplar on lower slopes to an association of oak–hickory on drier ridges.

Warbler "waves" can be expected in spring, and nearly twenty species have been found breeding. Birds found frequently during the breeding season are the Red–shouldered Hawk, Barred Owl, Acadian Flycatcher, Wood Thrush, American Redstart, Cerulean and Hooded warblers.

Roadsides– Kanawha State Forest has many open grassy areas for picnicking and ball fields. In their edges as well as in the tall trees along the roads may be found American

Robins and Goldfinches. Chipping Sparrows hop through the grass and American Redstarts flit in and out of the trees. Indigo Buntings are often seen in the unmown areas under power lines. Birds often heard singing in the edge habitat include Blue–winged and Parula warblers, Ovenbird, Carolina Wren, Gray Catbird, and Scarlet Tanager. Eastern Towhees call noisily from the ground. Wildflowers growing beside the roads attract Ruby–throated Hummingbirds.

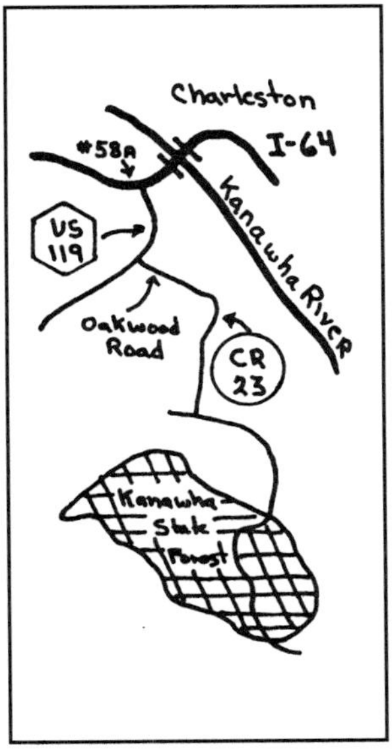

Pond– The pond at the entrance to the Kanawha State Forest is home to families of Mallards and Wood Ducks, with occasional domestic ducks and Belted Kingfisher. Because the area is open and grassy but bordered by the forest on two sides, it is also a good place to hear the calls of birds that live in the forest and to see those that dwell in edge habitat. Yellow Warblers, Red–eyed Vireos, Ovenbirds and Scarlet Tanagers can be heard. Phoebes are usually present around the picnic shelters. Eastern Screech–Owls are sometimes heard around dusk, Turkey Vultures and occasionally hawks are seen circling, and woodpeckers, particularly Northern Flickers, Red–bellied, and Pileated woodpeckers fly across the area.

CCC Snipe Trail/Swimming Pool Area– This is an excellent birding spot during the spring migration in March and April. The picnic area behind the swimming pool contains many old hardwood trees with little undergrowth. Further along the trail the forest becomes denser with a lot of evergreens such as laurel and hemlock. A stream runs through the hollow, and the gradual ascent of the trail along the slope allows views across the hollow into the mid–sections of trees. Very common spring migrants are the Kentucky, Black–throated Green, Cerulean, Black–and–white warblers, Ovenbird, Red–eyed, White–eyed, and

Yellow–throated vireos, and Wood and Hermit thrushes. The Worm–eating Warbler has been seen here. Acadian Flycatchers are obvious. Louisiana Waterthrushes call from the stream bed. Hooded Warblers become more common towards the end of the Snipe Trail where the **Mossy Rock Trail** begins, and will venture close to observers when foraging for food or defending territories.

Polly Hollow– This is another excellent birding area. Here (and Snipe Trail), one can dependably hear the Black–throated Green Warbler in spring. Along the lower **Polly Trail** and the gas well road are Ovenbirds, chickadees, titmice and Nuthatches, Red– and White–eyed vireos. As one walks further into the forest along the road, Wood Thrushes become more prevalent, and several may be heard singing at once. On the Polly Trail or **Beech Glen Trail**, one can see Scarlet Tanagers, Hooded and Cerulean warblers, especially along the ridge top.

Beech Fork Lake and State Park

Mike Griffith

Beech Fork Lake is located in Wayne County approximately twenty minutes south of Huntington. Take SR 10 off I–64 south, then turn right on CR 35 and right again on CR 54 . At CR 43, turn right to the park. The lake is managed by the U.S. Army Corps of Engineers as a flood control project. From November through March, the water level of the lake is 8.5 feet lower than it is during the summer months. **Beech Fork State Park** is located at the upper end of the main branch of the lake. Trail maps and other information can be found at both visitor centers located at the park entrance and the dam.

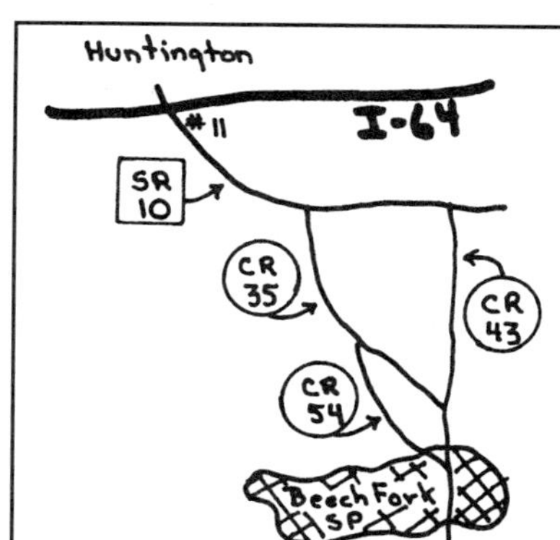

The forest surrounding the lake is mixed mesophytic hardwoods typical for this section of the state. There are also small, scattered stands of Virginia Pine. Oaks are usually predominant on the ridges and southern–facing hillsides.

The permanent residents list of both dam and state park areas includes Red–shouldered and Red–tailed hawks, American Kestrel, Great Horned, Barred and Eastern Screech–Owl, Red–bellied, Downy, Hairy, and Pileated woodpeckers, Northern Flicker, Carolina Chickadee, Tufted Titmouse, White–breasted Nuthatch, Carolina Wren, Eastern Bluebird, Eastern Towhee, and Eastern Meadowlark.

Common **winter visitors** include Golden–crowned Kinglet, Hermit Thrush, and White–throated Sparrow. Birds that are sometimes present in winter include Eastern Phoebe, Red–breasted Nuthatch (**Bowen Cemetery Road**), Brown Creeper, Winter Wren, Ruby–crowned Kinglet, Yellow–rumped Warbler, and Pine Warbler (Bowen Cemetery Road).

On the lake itself, in the spring, fall, and winter (usually after severe weather), a variety of ducks and other water birds occasionally show up. These include Common Loon, Ring–necked Duck, Lesser Scaup, Hooded and Red–breasted mergansers, Common and Forster's terns, and Ring–billed Gull. Osprey can regularly be found during spring and fall migration.

Any of the ridges around the lake can be very good for migrating passerines, particularly thrushes, warblers, and vireos.

Summer Residents – Birds that can be found anywhere in the woodlands and open fields around the lake include Yellow–billed and Black–billed cuckoos, Great Crested Flycatcher, Eastern Wood–Pewee, Acadian Flycatcher, Wood Thrush, White–eyed and Red–eyed vireos; Blue–winged, Yellow, Yellow–throated, Prairie, Worm–eating, Kentucky, and Hooded warblers, American Redstart, Ovenbird, Common Yellowthroat, Yellow–breasted Chat, Summer and Scarlet tanagers, and Chipping, Field, and Song sparrows.

Green Herons are commonly found and Canada Geese and Mallards nest at many locations around the lake. Killdeer nest near the dam and at the campgrounds. American Woodcock courtship displays can be observed near the state park visitor center between mid–February and March. Whip–poor–wills can be heard and Willow Flycatchers and Tree Swallows can be found in the marshy area near the visitor center. Eastern Kingbirds are common around the campgrounds. Cliff and Barn swallows nest at the state park under the bridge (Wayne County CR 17) that crosses the reservoir. Pine Warblers can be found on the Bowen Cemetery Road. Orchard Orioles occur around the campground.

Greenbottom Wildlife Management Area
Robert C. Byrd Locks and Dam
Huntington to McClintic WMA

Wendell Argabrite and Cynthia D. Ellis

Directions – Greenbottom Wildlife Management Area is 18 miles north of Huntington on SR 2 in Cabell County. (From Charleston, take I-64 to Milton and bear right at Exit 28. Sign is marked "Mount Olive" and route is known as Ashton-Upland Road. Drive approximately 20 miles to SR 2, turn left and go approximately 6 miles.)

Access and parking – Visitors may park at the most visible landmark on the GWMA – the historic Jenkins home. South of the home is a small parking area for a boardwalk; continuing south one finds impounded ponds and parking. North of the home is a short road to a dock on the Ohio River and parking is also available.

Habitat – The GWMA offers riparian fields, swamp, and small areas of hardwood forest. It borders a broad expanse of the Ohio River and includes several small backwaters. Its southern boundary is a two lane highway; this is flanked by a hillside ridge.

Precautions – There is hunting at GWMA beginning with the dove season in September and continuing through rabbit season in February. Safety is a secondary concern in this location as long as birders use courtesy and common sense; however, the presence of human activity has adversely affected the numbers of birds seen. There are both safety zones and good birding near the Jenkins home and the HQ building. There is no hunting on Sunday.

The only restroom is a portable unit at the Jenkins home.

Specialties – The impoundment ponds can host migrating waterfowl and shorebirds. The area between the Jenkins house and the Ohio River may be checked for breeding Common Moorhens, Sora and Virginia Rails, and Least Bittern. Look over the swampy area for Tree Swallows and Prothonotary Warblers. Willow Flycatchers and Blue Grosbeaks may be found in brushy areas. The large trees near the house are a reliable spot for Warbling Vireos. Breeding ducks, such as Wood Ducks, frequent the backwaters northeast of the house, as do varied duck species in migration. In winter one may see Northern Harrier, Rough-legged Hawks, and (in recent years) Short-eared Owls. January and February may bring Bald Eagles. More specifics are pro-

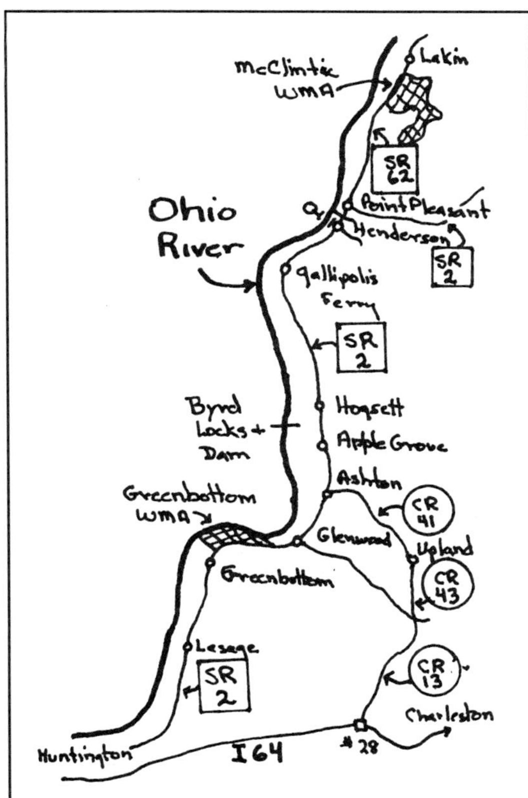

vided in the tour narrative.

Suggested birding route – Different seasons will prompt different routes for Greenbottom. Winter birders may wish to "car bird" from the roadside and berm of the raised highway north of the Jenkins home. Spring and fall birders will want to walk in any of the areas from parking spots noted above. Beginning birders or people on a limited schedule may wish to simply walk the boardwalk (damaged by floods in 1997, scheduled to be replaced) and "scope out" the ponds and the swamp.

Tour from Huntington to McClintic WMA – The following is a suggested tour from Huntington. It includes Greenbottom WMA, the Robert C. Byrd Locks and Dam, and the McClintic Wildlife Management Area.The entire tour covers 53 miles and visits the richest areas in this part of the state for a great variety of birds, particularly waterfowl and shorebirds.

The tour begins in downtown Huntington at Harris Riverfront Park. This is located near the Civic Center. From Veterans Memorial Boulevard, go through the floodwall to the boat ramp and marina. This has been a very good place to find gulls (including some records of Bonaparte's and Franklin's), terns (sometimes Black), and shorebirds. Check the large flocks of ducks resting here in winter.

Leaving the park, go to Fifth Avenue and travel east (about 2.8 miles) to Bridge Street in Guyandotte. From there turn left on Buffington Street, go through the floodwall, and follow the road to the mouth of the Guyandotte River. This area under the east end bridge to Ohio is good for terns, swans, and winter ducks (occasionally Oldsquaw).

Return to Bridge Street and turn left. This street becomes West Virginia SR 2. Travel north on SR 2 (about 15 miles) to the Greenbottom Wildlife Management Area.

At the southern edge of GWMA there is a parking area across the railroad tracks and a road (closed to traffic). Here are brushy fields where Willow Flycatchers and Blue Grosbeaks and sparrows (in winter) may be found. The fields may host Short-eared Owls in winter, as well as Harriers and Rough-legged Hawks. A pond at this location is great for egrets and herons; sometimes Little Blue or Night-Herons, Marsh Wrens and (sometimes) Sedge Wrens may be seen, along with swallows: Tree, Barn, Cliff, Bank, and Rough-winged. Continue walking on the road past the first pond until you come to another on the right. Here may be found migrating shorebirds such as Semipalmated, Baird's, and White-rumped Sandpipers (once a Black-necked Stilt).

Return to SR 2 and go .6 mi. to a road on the left. Cross the railroad tracks again and park at the headquarters building. Walk on the road and turn right to reach the boardwalk, or continue ontoward the river. This is another excellent area for any type of marsh bird. Virginia and Sora Rails nest; there was probable evidenc of King Rails nesting in 1997. Moorhens and Least Bitterns nest; American Bitterns migrate through. As in other areas nearby, one may find Blue Grosbeak, Prothonotary Warbler, Willow Flycatcher, Chat, and Common Yellowthroat. Sparrows include Song, White-crowned, Tree, Field, Swamp, and occasionally Vesper, Fox, and Lincoln's. Wood Ducks are commonly found. Kestrels nest here. Across the road Red-tailed Hawks nest; often seen and heard are Screech and Great Horned Owls.

Return to SR 2, go .2 miles, turn left and stop at the parking area near the large old white brick home. (This was the birthplace of Civil War General Albert Gallatin Jenkins and was build by his father.) The home faces the river. Large trees here are a reliable spot to find Warbling Vireo, Baltimore and Orchard Orioles. Beavers are active in the area. There is one record of Sandhill Cranes. A large expanse of swamp adjoins this site. Prothonotary Warblers and Swallow abound. Migrating swallow are easily viewed on spring evenings after rainstorms. Rails, herons, egrets, coots, grebes, moorhens, Wood Ducks, and Blue-winged Teal can be seen.

Continue north on SR 2 for 1 mile to a blocked road on the left. This is a great place for migrating warblers (including Northern and Louisiana Waterthrush) and blackbirds (including Rusty) in spring, and for Prothonotary Warbler in summer.

Go north on SR 2 for .7 mile to a road on the left; follow it .5 mile to a boat ramp. Along this road a variety of sparrows may be seen; in fall look for Savannahs.

Among other birds are Horned Larks and hawks and owls mentioned earlier. The boat ramp is especially good for views of terns; also gulls, shorebirds, and ducks. Backtracking up the boat ramp road is a parking area with good brush habitat for wrens. Dickcissels have been seen here.

Return to SR 2 and travel 4.6 mi. north to the Ashton Post Office. The state's only Hoary Redpoll was seen near here. Sometimes Peregrine Falcons can be seen.

Continue on SR 2 for 2.2 mi. looking for a small pond on the left as you approach a factory. This area is good for hawks: Red-tailed, Harriers, Sharp-shinned, Rough-legged, and sometimes Merlin and Peregrine Falcon. There are single records at this pond for Glossy Ibis and Cattle Egret. Waterfowl and shorebirds can be numerous. Check in late summer for Ruddy Turnstone and Stilt Sandpiper.

Go 1.5 mi. north to the access road for the Robert C. Byrd Locks and Dam. This site rivals, indeed surpasses the Greenbottom area for great birding opportunities. Bald Eagles may be seen here in late winter; look also for Short-eared Owls, Harriers, Rough-legged Hawks, Osprey, and resident birds of prey. The fields that flank the access road are good for Horned Larks, pipits, and (once) Snow Buntings. Fifty yards along this road, on the left, is a gully where dickcissels nested in 1997. One can usually find nesting Grasshopper and Savannah Sparrows. In the fall, the mowed areas may host Golden and Black-bellied Plover, and Upland and Baird's Sandpiper.

Continue on this access road and bear left .4 mi. to a parking area. Pass through it, still bearing left, to a gravel road. At midpoint is a portable toilet. Birders may choose to walk up the dike behind that and then walk in to a wet zone. Some waterfowl sightings from this place include Double-crested Cormorants, Ross's Goose, and Buff-breasted Sandpiper. From the gravel road on the right one sees the waters from the dam. This spot offers a sheltered nook for waterfowl in winter.

Return to the parking area and go .4 mi. to the Visitor's Center. Stay within the fences! Walk left of the Center to view the upriver area. This has been a terrifically productive area for waterfowl. Look for gulls, terns, loons, and ducks – especially diving ducks. You may see Scaup, Goldeneye, Ruddy, Bufflehead, and Canvasback. Gulls, terns, and even shorebirds may be seen resting on the concrete pier.

Across the inlet is an island. This has been used as a fall

staging area for shorebirds such as dowitchers and Buff-bellied Sandpiper. There is one record of Red-necked Grebe.

Rest rooms are open in the Visitor's Center on week days.

Return to SR 2 and continue north. There are a number of places where you may pull off onto side roads and view small ponds or wet areas. Continue to the city of Point Pleasant. Here the Kanawha River joins the Ohio.

Turn right at the first light from the bridge to Krodel Park. A pond often has waterfowl, including mergansers in winter. Return to SR 2 and go through Point Pleasant, driving north on SR 62 about 6.3 mi. to the McClintic Wildlife Management Area. This land was once used as a TNT plant and several "igloos" or domed storage areas from the World War II era can be seen. Since then it has been managed for hunting and fishing.

There are several access roads into McClintic from SR 62 and numerous roads criss-cross it. Check the fishing ponds for waterfowl; fields and brushy areas for warblers and sparrows. You may hear Bobwhite and see Wild Turkeys.

Small and casual restaurants are available in Point Pleasant. There are many restaurants in Huntington. Motel accommodations in Huntington are mainly located on US 60 East.

6. The Mid–Ohio Valley Region

Jane Whitaker

This western region of West Virginia is bordered by the **Ohio River** and runs from approximately Ravenwood at the south to about Sistersville to the north. It has low elevations of 500 to 600 feet along the river, but the hills rise to just under 2000 feet before entering the Mountain Lakes Region to the east. Along this section of the Ohio River you will find some areas of exceptionally steep wooded bluffs rising from the river's edge, while other areas adjacent to the Ohio have broad valleys. These stretch eastward along the tributaries towards the mountains. The flatter areas are mostly farmland with habitat suitable for birds of open spaces. Along the river are several large power and chemical complexes, which often have standing, shallow water on their premises that attract migrating shorebirds.

There are several locks and dams along the Ohio that attract gulls, cormorants, and migrating waterfowl. The **Ohio River Islands Wildlife Refuge** protects several islands from Paden City downstream to Parkersburg.

Geologically, the Mid–Ohio Valley is composed of primarily Pennsylvanian–Permian shales and sandstones which supply numerous oil and natural gas wells. There is little, if any, coal production.

The steeper hillsides are second growth hardwood forest with some stands of conifers. This part of West Virginia is well populated due to the industry adjacent to the Ohio River and the availability of cheap commercial transportation from Pittsburgh to New Orleans via river barges. The glass manufacturing industry has been active here for over a century.

The **North Bend Rail Trail** runs from near Parkersburg east for 60 miles. There is access from several points along US 50. The trail gives outdoor enthusiasts easy access to some of the more remote and less populated areas of the region. **Cedar Creek** and **North Bend** are the only state parks in the Mid–Ohio Valley region but there are several wildlife management areas that are open to the public that offer good birding habitat.

Ohio River – Wood and Pleasants Counties

Dick and Jeanette Esker

Washington Bottom – The Ohio River valley southwest of Parkersburg, WV was once owned by George Washington, hence the name. This area is primarily farm fields surrounded by industrial sites with some woods on the edge. There is access to the Ohio River.

Take SR 68 south from Parkersburg about three miles (four lane road) to the first stop light outside Parkersburg. Proceed straight ahead at the light onto SR 892, also known as DuPont Road. (SR 68 goes off to the left at the light). Proceed to where the road drops down into the valley. On the left at 2.8 miles from the light just as the trees along the road end, is an entrance to the landfill. Just to the left of the gate is a small parking lot. This is the beginning of the **DuPont Wildlife Enhancement Committee Nature Trail**, open to the public. The trail is about a mile long, rising gradually to the top of the ridge overlooking the river and continue back down a moderately steep path to the parking lot. The beginning of the trail is open with brushy areas and has a small pond. The remainder leads through mature hardwood woods. Warblers are plentiful during spring migration. Blue–winged and Yellow warblers nest in the area, as does a Red–tailed Hawk. The usual woodland birds can be found. A flock of two dozen Wild Turkeys is often seen.

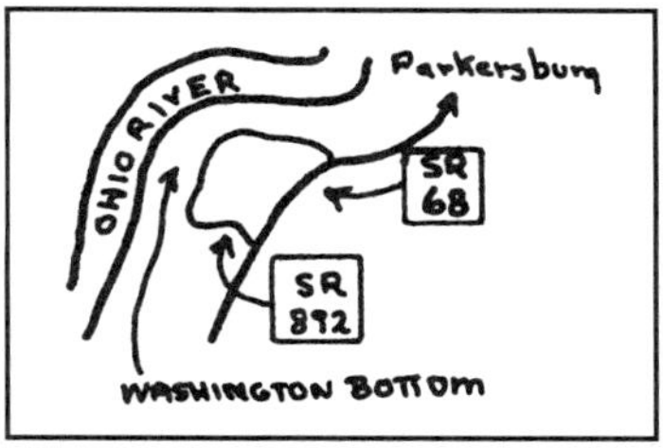

Return to SR 892 and proceed another 1.4 miles, past the DuPont and GE plants. The fields beyond GE are a favorite location for Northern Harriers in the winter, and occasionally the Rough–legged Hawk. Watch for Kestrels on the wires and posts. Red–tailed Hawk are often seen.

Look for a water tower with "Foster" painted on the side. Take the road just before the water tower to access the

Ohio River. There is a public access area at the site of an old lock and dam and is a good location to scan the river during waterfowl migration. The small grove of trees and brush on the right 0.5 miles down the road attracts White crowned, White–throat, Field, Tree and Song sparrows in winter. Watch all along the access road for unusual species because this area seems to attract many of the migrants. Bobolinks, Lincoln and Savannah sparrows have been seen.
Return to SR 892. Turn left to retrace the route back to the stop light at SR 68.

McDonough Wildlife Refuge – The McDonough family donated hundreds of acres to the City of Vienna as part of the park system. It is primarily wooded with some fields and open areas and a couple of ponds. It has excellent trails that provide access to most of the area.

The Wildlife Refuge can be accessed from Vienna, WV or Parkersburg, WV via Rosemar Road. From Parkersburg, take SR 68 north. At the edge of Parkersburg, Rosemar Road intersects SR 68 at a traffic light. (If you are coming into Parkersburg from the north, Rosemar Road is 3.0 miles south of the I–77 and SR 68 intersection). Turn left (west) onto Rosemar Road and proceed 2.2 miles. Look for the entrance sign on the right, across the road from several ball fields.

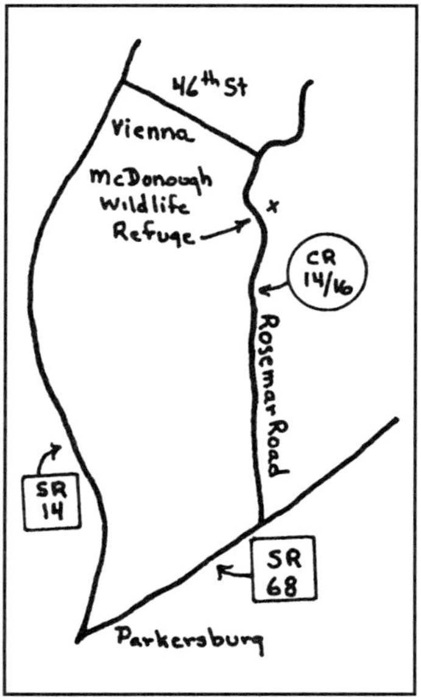

From Vienna, take 46th Street to where it T's with Rosemar Road. Turn right onto Rosemar Road and proceed 0.4 miles. The entrance is on the left.

Drive along the entrance road to the parking lot at the top of the hill (about 0.1 mile). The trail that goes up the steps at the far right of the parking lot follows a ridge. This is a good area for spring migrating warblers. This trail continues on through the wooded area and circles back to the parking lot, is 2.1 miles long, and there is a cutoff that covers about half the distance.

The trail near the middle of the parking lot that goes downhill leads to the ponds. The Blue–winged Warbler nests in this area. The Yellow–throated Warbler is often found in the Sycamores along the creek that drain into the ponds from the right.

Blennerhassett Island –The land surrounding the Mansion has large Sycamores, Poplar and Cottonwood trees and a large grove of Walnut trees. A wide variety of warblers can be seen during migration. Baltimore Orioles, Yellow and Yellow–throated warblers nest on the island and are often seen around the Mansion. Prothonotary Warblers nest on the island but are usually difficult to find. There is always an abundance of House Wrens during the summer.

The island is accessible only by boat. A paddlewheel ferries visitors to the island from May through October for a fee. Hours are typically from 10:30 am to 5:30 pm. The boat leaves from Point Park at the confluence of the Little Kanawha and the Ohio River, at the foot of Second Street in downtown Parkersburg, outside the flood wall.

Mountwood Park – Mountwood Park is a lovely wooded area with a lake, hiking trails and picnic grounds. It is host to the typical woodland birds of the area. There are some open areas where a variety of warblers can be found during spring migration.

. Mountwood Park is east of Parkersburg along US 50, just off US 50 on Volcano Road. It is 11.8 miles east of the I 77 and US 50 intersection. This was oil–boom country in the early days of the oil industry, hence the name Volcano for the oil gushers.

Just after turning off US 50 on Volcano Road, the lake is visible. The first turn–off (0.1 mile) leads to a parking lot at the backside of the lake and one end of the trail around the lake. About 0.5 miles farther along Volcano Road is another parking lot and boat ramp. This is also the other end of the lake trail. Volcano Road proceeds through the middle of the park, with picnic grounds along both

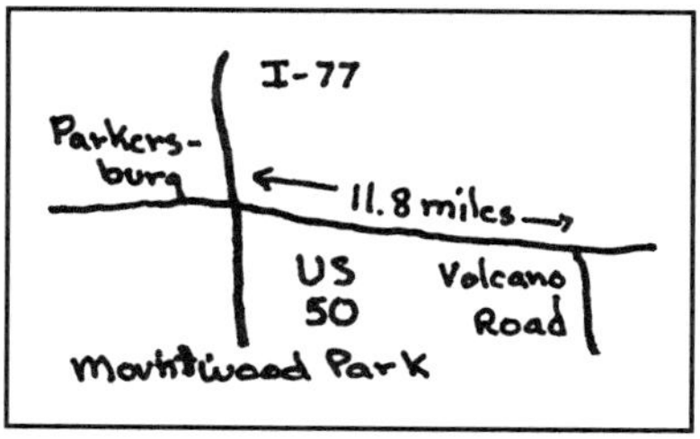

sides of the road for about a mile. The woodline around the picnic grounds often provides good birding. After the picnic area Volcano Road winds up a hill. The first turnoff to the left (0.4 miles) leads to the White Oak picnic area, another good birding spot. This area is a little farther (0.7 miles) from the main road and is therefore a little quieter. Return to Volcano Road and continue another 0.2 miles. The entrance to the Nature Center is on the right. The gate across the entrance may be locked. It is about 0.1 miles walk back to a trail that follows a power line just beyond the buildings and is a good area for warblers. Return to US 50 via Volcano Road.

Belleville Locks and Dam –The locks and dams along the Ohio River typically offer public access to the river for fishing and an observation tower for watching barges pass through the locks, and offer good viewing for birds along the river. The turbulent water downstream of the dams attracts many waterfowl.

Belleville Locks and Dam are south of Parkersburg, WV on SR 68. It is 12.2 miles from the intersection of SR 68 and DuPont Road (SR 892) to the dam access road. There are signs directing you to the fishing area. A hydroelectric facility is being constructed next to the dam. The access road has been relocated because of the construction and currently goes to the left just after crossing the railroad tracks a short distance from SR 68. This is a good area to look for gulls and winter water-fowl. Cliff Swallows nest on the concrete dam structure under the ledges near the top.

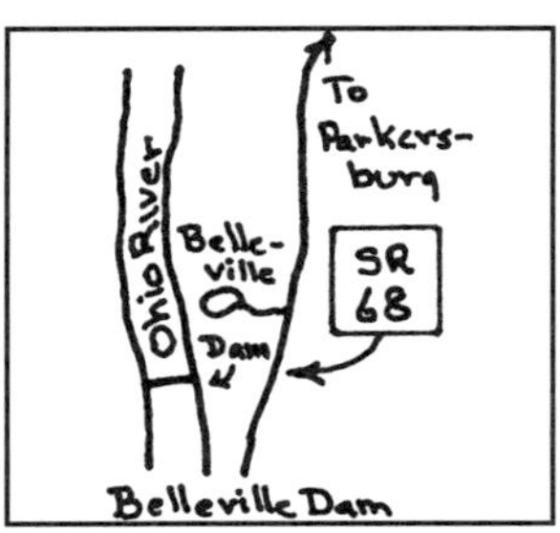

French Creek Backwaters – A backwater is created where French Creek enters the Ohio River. This area is just north of Belmont, WV, on SR 2 (13.6 miles north of the I 77 and SR 2 intersection north of Parkersburg). Best access is at a small boat ramp and parking lot at a public access area along SR 2 just south of the bridge that crosses French Creek. These waters attract a variety of birds, especially winter ducks and gulls. Bald Eagles and Osprey are often seen. Cliff Swallows nest under the SR 2 and railroad bridges.

The backwaters extend up French Creek for a couple of miles. To access the back side, return to Belmont via SR 2 (0.3 miles). Turn left onto Main Street. This is also Henry Camp Road (CR 1). The road makes a sharp left turn after 0.3 miles and continues out into the country. The backwater is visible on the left from the road at 1.0 to 1.3 miles. The shoulders are narrow so there is no room to park until about the 1.2 miles section.

Willow Island Locks and Dam – There are also Ohio River locks and dam near Belmont, WV. The access road is 4.5 miles south of French Creek, or 9.1 miles north of the I 77 and SR 2 intersection north of Parkersburg. There is a sign along SR 2 at the road that says, "Fishing Access Willow Island Locks and Dam." It is a little difficult to see since it is up on a bit of a bank. Turn west off SR 2 onto the access road. It is 0.7 miles to the dam. The river can be scanned for waterfowl from the bank. The concrete slab at the end of the dam offers a good view up and down the river. Also, step over the guardrail at the end of the parking lot and walk over to the bank of the adjacent cove (only about 100 feet from the parking lot). This is a good place to look for ducks.

The Ohio side of the dam has an observation tower. The most convenient way to get to the Ohio side is via the bridge at St. Marys, WV, just 6.8 miles north of the dam access road, on SR 2 (2.3 miles north of French Creek). Cross the bridge to Ohio. On the Ohio side the bridge ramp makes a U and ends at SR 7. Turn right onto SR 7 south and travel 6.5 miles to the dam.

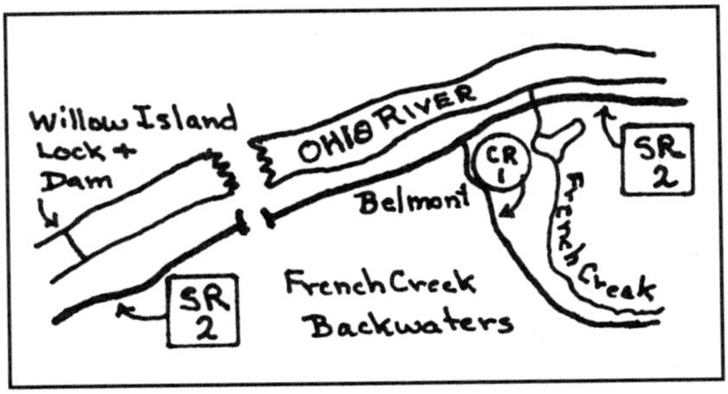

While on the Ohio side, another backwater area to check out is Newell's Run, just across the Ohio River from French Creek. Newell's Run begins 2.8 miles south of the St. Marys bridge. There is a side road (Newport Township Road 19) that goes around the back side of the backwater. Road 19

intersects with Township Road 443 that returns to SR 7.

Ohio River Islands National Wildlife Refuge – A number of islands in the Ohio River from Pennsylvania to Kentucky have been purchased and set aside as a National Wildlife Refuge. Some islands of interest in the Mid–Ohio Valley area are:

Grape Island, which is about three miles north of St. Marys, WV., has a Great Blue Heron rookery at the north end of the island. Over 100 nests have been counted. The best viewing spot is from the Ohio side, 3.6 miles north of the St. Marys bridge on Ohio SR 7.

Muskingham Island is about seven miles north of Parkersburg. This island is accessible only by boat, is a good spot to see warblers during the spring migration if you can find a way to get out there.

North Bend Rail Trail

Jane Whitaker

Ritchie County is located in west–central West Virginia and is easily accessible from I–77 or I–79. It is traversed by US 50, which links the respective interstates at Parkersburg and Clarksburg. The **North Bend Rail Trail**, a 60–mile stretch of the old B&O railroad, is currently being upgraded to a state park. It starts in Wood Co. at Walker, runs through Ritchie and Doddridge counties, and ends near Clarksburg in Harrison County. It runs east–west and is easily accessible from US 50 at several locations.

A variety of habitats, including remote woodland, active farms, small towns, brushy overgrowth, and even a glass factory are intersected by the rail trail. The walking is easy. The only unusual requirement is a good flashlight as there are numerous tunnels, one almost a half mile long, which cannot be easily circumnavigated.

The flora is enhanced by the railroad right–of–way with many stone outcrops. Lichens, ferns, and a variety of angiosperms, such as purple–flowering raspberry, can be found along the trail. You can expect to see Brown Thrasher, Scarlet Tanager, a variety of warblers, Yellow–throated Vireo, Red–tailed Hawk, and an abundance of woodpeckers.

A good starting point is at **North Bend State Park** at Cairo. Approaching from the west, take SR 31 south off of US 50 to Cairo and then left on CR 14 to the park. From the east, at Ellenboro and US 50, take SR 16 south to Harrisville and then CR 5 to the park. Drop by the lodge and pick up a map to the rail trail. It will show you access points and tunnel locations. The park also has bird, flora, and fauna lists which will give you a good idea of what you may see while visiting Ritchie County. On weekends the park offers a horse and buggy ride on the trail (five mile trip) to Ellenboro.

A productive auto excursion of the rail trail area is as follows: From the North Bend Lodge, drive 0.5 mi. back to the park entrance and turn left towards Harrisville. Upon reaching Harrisville follow Main Street (SR 16) and continue towards Ellenboro and US 50. At the intersection of US 50 turn right onto 50 east to Pennsboro (4.5 mi.). Stay on the four–lane highway and pass the first exit for Pennsboro. Continue 1.5 mi. where you will see a sign for Spring Run on your right. If you get to SR 74 you have gone too far. Make a right turn onto the Spring Run Road. This is a little traveled, poorly–maintained road, but the first 0.2 mi is gravelled and very accessible. Drive the 0.2 mi to the bottom of the hill and park. Look for Killdeer in the large field on your right. Red–headed Woodpeckers will probably be in the large trees on either side of the road. You should be able to see Eastern Meadowlarks on some of the electric lines and you should see or hear Red–winged Blackbirds, Song and Field sparrows, Eastern Towhees, Bluebirds and possibly Kestrels.

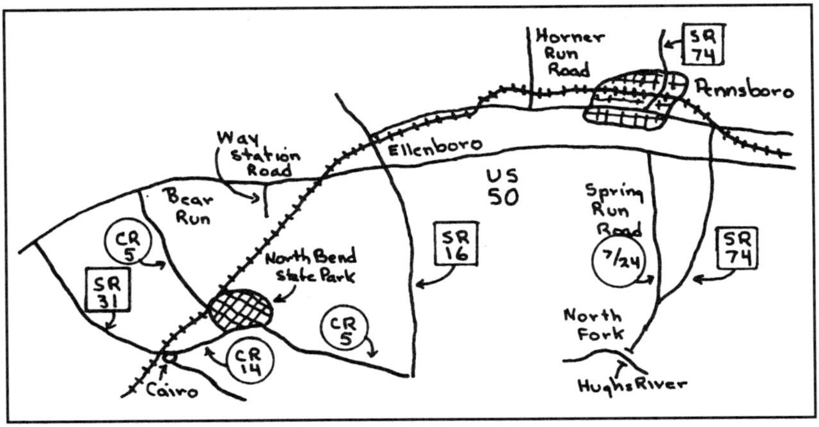

Walk along the road, being sure to scan the overgrown field to your right as well as the trees on the hill above the hollow. In summer, you can expect to see or hear Brown Thrashers, Catbirds, Mockingbirds, Field and Song sparrows, Blue–gray Gnatcatchers, Bluebirds, Goldfinches, Prairie Warblers, Common Yellowthroats, Eastern Towhee, Red–tailed Hawks and an abundance of woodpeckers; Red–headed, Red–bellied, Downy, Flicker, and Pileated. In winter you may find Yellow–bellied Sapsuckers, too.

Both Scarlet and Summer tanagers may be seen along the road as well as Phoebes and Acadian Flycatchers. There is no human habitation along this 2–mile road and it lends itself to a peaceful couple of hours of undisturbed birding. The roadside foliage is never cut, so there is an abundance of flora – wintergreen, spring beauties, bluets in spring and goldenrods, fall asters and ferns later in the year.

Side trip – A right turn onto SR 74 south will lead to the **North Fork of Hughes River** in 3.0 mi. Instead of crossing the river, bear right and park along the secondary road. A short walk of 0.1 mi. to the sharp curve above the river bluff should result in more sightings of Red–headed Woodpeckers. In spring a myriad of wild flowers can be observed along the road bank, dwarf iris, violets, bluets, saxafrage, and trillium are only a few.

The 1.0 mi. walk along the old jeep trail which follows the river will reap a number of floral and fauna finds. The road traverses a wooded cliff with interesting rock formations above on the right. There probable are an abundance of fence lizards among the rocks and frogs in the marshy area adjacent to the river. Belted Kingfishers, Field Sparrows, Blue–gray Gnatcatchers and Yellow Warblers are common. Spring wildflowers are plentiful – trillium, wild ginger and white, yellow and purple violets. Both species of spring beauty, bloodroot, and squawroot can also be found.

After 0.5 mi. the road becomes very muddy and if there has been much rain a circuit to the right through the woods is suggested. In the spring be sure to look for mating salamanders in the small pool to the right of the road. The gelatinous lichen, *Leptogium burgdorfii*, can be found among the moss–covered rocks here.

The small meadow on the left adjacent to the river is

easily accessible but ends in a mass of impenetrable briars further downstream. However, the meadow should be birded for sparrows, gnatcatchers and cardinals. If you proceed to the small ford you are at **Granny's Run**. The river bank which is visible from the far side of the run houses several Belted Kingfisher nests and the sandbar beneath the bank is always awash with the shells of freshwater mussels. Live specimens can be found in the riffles.

There is an abundance of wildflowers here in the spring – hepatica, anemones, trillium, wild geranium, violets and bloodroot. A variety of lichens are always present along the road bank. In late fall and winter you can find White–throated Sparrows in the brambles throughout the meadow. Red–tailed and Sharp–shinned hawks reside in this area.

Unfortunately, this bountiful area may be flooded by a proposed recreational dam at North Bend State Park. It is one of the remarkable sites that has become more prolific and diverse in the past 50 years and it will be tragic if it is destroyed.

When you return to your car, backtrack on SR 74 to US 50. Continue on SR 74 into Pennsboro, 0.5 mi. At the stop sign, the rail trail is directly in front of you but it immediately enters No. 7 tunnel. Turn left onto old Route 50 (two lanes), drive 0.3 mi. to the railroad crossing on your right and park. You may walk back towards the tunnel along the trail. This is a favorite place for Gray Catbirds, Northern Cardinals and Song Sparrows. As you approach the tunnel you will see cattails and other water–loving plants. This particular tunnel is quite wet and may not be possible to traverse. The rail trail runs west through woodland for about 1.5 mi. then opens onto pasture land. If the tunnel is not accessible and you wish to bird this area drive to the top of the tunnel, park at the abandoned store, and walk east on the paved road until you can access the trail. Return to your car and continue west through Pennsboro on old US 50.

Side trip – **Pennsboro City Park and Tracy Lake**. Just past the Pontiac dealership (0.65 mi. from the R.R. crossing before No. 7 tunnel) on old US 50 turn right onto SR 74 and follow it until it crosses the rail trail at the old depot. Turn left just past the depot and follow the Rose Hill Road sign. After 0.1 mi. you will turn right (note the fishing sign) onto Park Drive and climb a long, steep hill. Drive 1.4 mi. from the top of the hill looking for Indigo Bunting, Pileated Woodpecker, and Wild Turkey as you go. Pass the first turnoff on your left (1.4 mi.) and proceed 0.1 mi. to the next one. Turn right and take the road going down a steep hill to Tracy Lake.

Tracy Lake is a 2 + acre preserve owned by the city of Pennsboro. It is stocked with trout yearly and is home to Green Heron and Wood Duck. Occasionally Northern Harriers are found there. Yellow–breasted Chat nest in the brushy thickets at the shallow end of the lake. Look for Pearl Crescents, Wood Nymphs, swallowtails and Red–spotted Purple butterflies as well. Beavers are in the upper pond but are quite elusive. The wooded hillsides have Carolina Chickadee, Eastern Towhee, Tufted Titmouse, Pileated Woodpecker, Scarlet Tanager, Cerulean and Hooded warblers; however, this area is notorious for harboring Copperhead Snakes so be careful. The brushy field at the shallow end of the lake has nesting Redstart, Field Sparrow and Blue–winged Warbler. Wild Turkey are seen or heard in this area and Yellow–throated Warblers frequent the trees near the small pond above Tracy Lake.

Return to old US 50 in Pennsboro and continue west 1.75 mi. to the intersection at the racetrack. Bear right and you are still on old US 50. Proceed 0.2 mi. past the shopping area and turn right onto Horner Run Road. Drive 0.1 mi. and park by the rail trail. You will be about 0.05 mi. from No. 8 tunnel on your right. This is a good tunnel to investigate as it is quite dry and short enough that you can enter without a flashlight. The rock ledges along the trail harbor ferns, lichens and wild flowers. This is a good place to look for White–eyed Vireos.

Return to old US 50, turn right and continue paralleling the rail trail. The trail is visible from the road for the next 4 mi. and can be accessed at several points. The habitat in this area is primarily brushy meadows with Indigo Bunting, sparrows and goldfinches. Butterflies are numerous due to the wildflowers.

In 3.3 mi. you may stop at the small drive–in restaurant and access the trail. If you walk about 0.2 mi. west you will be in the midst of a glass factory! You can see the blast furnace belching fire and molten glass only a few yards from where you stand.

Continue driving old US 50 into Ellenboro (1 mi.) and turn left onto SR 16. Here you may return to North Bend Park on SR 16 or you may access the 4 lane US 50 and proceed 1.0 mile west where you may turn left and enter Way Station Road. Park here or turn right and drive 0.2 mi. to the trail and park. Walk the rail trail west. You are now in a woodland environment and entering one of the more remote sections of the rail trail.

The three mile walk on this section of the trail in uninhabited, with wooded steep rocky ridges. Mixed pine and hardwood forest is on both sides of the elevated rail bed and a small stream with occasional back waters is adjacent to the trail. There are 3 tunnels along this section of the trail so a flashlight is necessary. See page 136 for additional information concerning the Cerulean Warbler.

Flycatchers, Yellow–throated and Black–and–white warblers are some of the resident species. In spring, woodland wild flowers such as trillium and spring beauty are plentiful. Butterflies abound from early spring until late fall.

One steep hillside about 0.3 mi. east of No. 12 tunnel is covered with Dutchman's breeches in late April only to be replaced by maidenhair fern later is the summer. Equisetum is common.

Whip–poor–will
Virginia Cronenberger

Whip–poor–wills can still be heard in an area **near North Bend State Park** on and around the farm of a BBC member. From Ellenboro, take US 50 west for about 10 miles. Then turn south on SR 31 to Cairo. Continue through town and turn right on CR 15. Drive about 5 miles and turn left onto Racy–Mellin Road (CR 17), (Note the Blankenship–Sammons farm at the intersection). Go another 1.7 miles to Racy noting the old Racy General Store and North Fork Baptist Church at this intersection. Continue on the paved road, CR 17.

Alternately from Parkersburg, go south on I–77 and at exit 174, turn east onto SR 47. Proceed to Cisco and turn left onto Cairo–Cisco Road (CR 15). Go 2.3 miles and turn right on Racy–Mellin Road (CR 17), and continue into Racy. Continue on CR 17.

From this point on, begin to listen for the song of the Whip–poor–wills at dawn and dusk. Continue another mile to the white one story house with a porch across the front on the hillside on the left side of the road. Ken and Virginia Cronenberger live here and are willing to receive phone calls (304–628–3382) to give current information about the Whip–poor–wills.

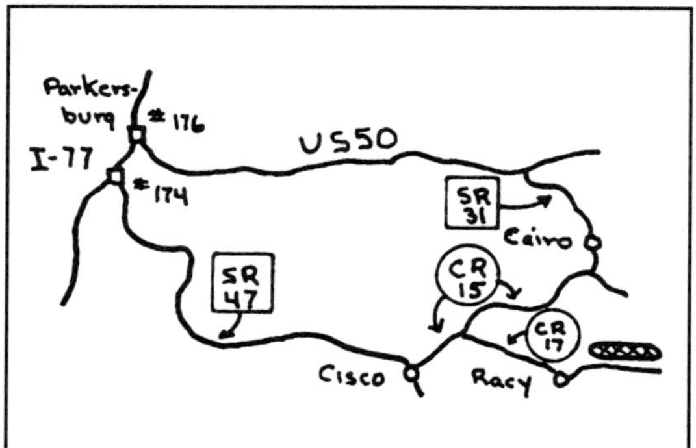

7. Northern Panhandle

Greg Eddy

This area of West Virginia consists of Hancock, Brooke, Ohio, Marshall and Wetzel counties. The local relief in Wetzel County is about 800 feet and the topography consists of very steep slopes and narrow valleys. The relief decreases to about 500 feet in Hancock County and the slopes and valleys become more subdued.

There are only a few special places for birding in this part of West Virginia. Although West Virginia is not on a waterfowl migration flyway, the **Ohio River** can be rewarding at times. In the winter when Lake Erie freezes or has very bad storms, some of the gulls and ducks that winter there temporarily leave the lake and come south to the Ohio River. At those times hundreds of Ring–billed and Herring Gulls and rafts of hundreds of Common Goldeneyes and Lesser Scaup can be seen. Mergansers and other duck species are also seen, but in much smaller numbers. One of the best places to observe waterfowl on the river is from **Pike Island Dam** north approximately three miles to the marina at Rayland, Ohio, called Warrenton Inlet. The marina is in Ohio, but the river belongs to West Virginia. About a half mile north of the inlet an Osprey nest has been active in recent years.

A species of note is the Yellow–throated Warbler. They were a very rare species in this area only 20 years ago but are now found along nearly every stream. They were first seen at Viola on Big Wheeling Creek about seven air miles from the river, a marvelous area for spring wild flowers and early summer birding. Another species of note is the Cerulean Warbler, which seem to be disappearing from some of their historical territory. Singing Male Population Studies by the Brooks Bird Club show that their numbers have declined by about 50 % over the last 30 years.

Great Blue Herons are common. They can be seen along the Ohio River, every creek and visiting every lake and pond. Several heron rookeries are present the largest of which is on **Fish Creek Island** in the Ohio River about seven miles south of Moundsville. Possibly a hundred nests are present here. A rookery of more than 30 nests is present on a hill north of **Big Wheeling Creek** one mile down stream of Viola.

A good side trip in the Wellsburg area is a trip to **Funk's Shangri–La**. Funk's Shangri–La Road, CR 16, is 5.1 miles east of Wellsburg on SR 27. Turn right, south, bear left down hill for 0.3 miles to two small ponds. Mr. and Mrs. Funk feed over 3,500 pounds of feed a year to the ducks and geese that visit the ponds. This attracts many of the local Canada Geese and Mallards, and a few migrants are usually present. This is easily the best spot to see Wood Duck during migration.

Oglebay Park
Greg Eddy

Oglebay Park, located in **Wheeling**, is one of the largest municipal parks in the country. Follow the signs to Oglebay Park on I–70 by exiting on SR 88 north and continuing to the park. The park covers 1500 acres and contains many of the usual park facilities, such as golf courses, cabins, a swimming pool, small zoo, lodging etc. It also contains the **Brooks Nature Center** and associated nature trails. The nature center has interesting exhibits and a comprehensive nature oriented–book and gift shop, adjacent to which is a beautiful Butterfly Garden and the **Dorothy Broemsen Hummingbird Garden**.

The best spring and summer birding in the park is on the nature trails leading from the nature center and other locations, i.e., **Caddy Camp**, **Serpentine Trail** and **Pogue Run**. The staff will give directions to these and other good locations for observing spring migrants and summer residents of the mature hardwood forest. These can include Cerulean and Kentucky warblers and will include Wood Thrush, Acadian Flycatcher and Ovenbird.

The best birding in the park during the winter is at the Nature Center's **feeders**. Every seed and suet eating winter resident can be seen including Carolina Chickadee and in recent years, Black–capped Chickadee, Tufted Titmouse, White–breasted Nuthatch, Carolina Wren, Northern Cardinal, Dark–eyed Junco, Tree, Song and White–throated sparrows. Every resident woodpecker can be seen at the feeders. This is the only location the author has seen more than two Pileated Woodpeckers at the same time. Pine Siskins have been regulars at the feeders even when they haven't been seen at other locations in the area.

The Nature Center has a fine schedule of nature activities for the public. This includes several nature walks and especially spring and early summer bird walks. Contact the nature center at (304) 242 –6855.

Weirton and Vicinity

Bob Rine and Bill Murray

Although Weirton is a steel mill town, there is good birding nearby. It is not far from the mills to the trills. Two state parks and a county park are 30 minutes or less from Weirton. **Tomlinson Run State Park** is about 15 miles north of Weirton, via SR 2 and SR 8. The park has two main sections — a recreation area with swimming pool, picnic tables, three shelters, miniature golf, tennis courts, two hiking trails, a group camp and a 30–acre lake with fishing and boating. A camp ground, with washrooms and a store, plus four fishing ponds, are also available.

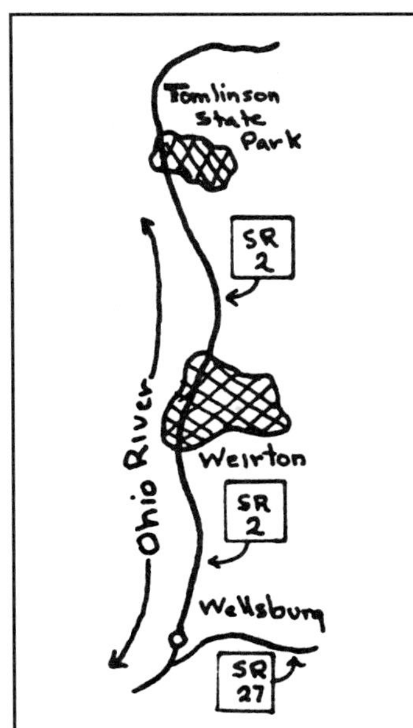

The other section is considered a wilderness area. A small site is used for Boy Scout tent camping but the rest is undeveloped and has two fine trails, **Laurel and White Oak trails**. The park covers 1400 acres and extends about 2.75 miles from the entrance to near the Ohio River to the west. At the lake, Canada geese, Mallards and other ducks mingle with sandpipers and other waders during the Spring and Fall migrations. Scarlet Tanagers are among the more colorful nesters and the whole range of wood warblers is seen during the migration periods.

Tomlinson Run is a unique stream when compared to others in West Virginia's northern panhandle. It is in a narrow valley which makes the plant life interesting and similar to that in the Canadian life zone. Canadian yew, Canada mayflower, rhododendron and mountain laurel en masse, trillium, red and white pines and hemlock, plus numerous flowering plants and ferns are found along the trails. In recent years, there has been a decrease in the number of individual birds and species in the park, but it remains a rewarding experience for the visitor.

Brooke Hills Park, a county park, is 20 miles south of Weirton, via SR 2 to Wellsburg, then SR 27 five miles east to the park. The best birding is past the golf course and summer theater barn to the woods beyond. Many warblers are there in season, including nesting Yellow Warblers and Scarlet Tanagers, among others; and Green Herons and Red–winged Blackbirds are seen at a small pond.

A short but rewarding trail is on **Weirton Heights**, on Colliers Way a half–mile past the Weirton Medical Center. A metal gate on the right leads the birder down a fairly open trail, where White–eyed and Red–eyed vireos, Indigo Buntings, Carolina Wrens, American Goldfinches and Song and Field sparrows can be seen. Rose–breasted Grosbeaks and several species of warbler pass through in Spring and Fall. Horned Larks and Grasshopper Sparrows have been seen in the large open fields opposite the trail.

8. Mountaineer Country

LeJay Graffious

The Mountaineer Country region of northern West Virginia is located between the state's northern and eastern panhandles. This region is the most diverse in the state due to its significant elevation differences, ranging from over 3,000 feet in southern Preston County to under 1,000 feet in parts of Monongalia and Doddridge County.

The Monongahela River valley forms the central basin of the region and divides the region's two main physiographic features, the Allegheny Plateau and the Allegheny Mountains. The Allegheny Plateau to the west is an area of rolling foothills composed of central hardwoods, primarily oak and hickory with some sycamore, elm and black walnut. The Allegheny Mountains rise to the east with northern hardwoods, a mixture of oaks, hickories, maples, birches, ash, basswood, cherry, black locust, tulip poplar, walnut, beech and others. The eastern–most border is the divide of the Ohio River and Potomac River watersheds.

The wide variety of ecological niches found in this region's diverse elevations includes high mountain bogs, forests of various ages, abundant farmland, rivers and impoundments. The chickadee signifies the division of central and northern forest types with Black–capped breeding in the eastern quarter of the region and Carolina nesting in the region's western parts.

The breeding season is the most predictable time for rewarding birding in the Mountaineer region. During the spring, migration is most apparent in the lowlands. Look for migrating passerines from mid–April through May. The late fall and winter migration of waterfowl can be observed from early November through March. During peaks in waterfowl migration, weather fronts often stall the birds and large numbers may be seen on ponds, rivers, and impoundments.

Besides the hot spot reports that follow, you may also want to visit **Cheat Lake** (exit 10 off I-68), **Prickett's Fort State Park, Valley Falls State Park, Tygart Lake State Park,** and **Tibb's Run Reservoir** (east of Morgantown between Tyrone and Dellsow off CR 75). For more information call LeJay Graffious at 1–304–379–4492.

Core Arboretum

Sally Stebbins

The **Core Arboretum** on the **Evansdale Campus of West Virginia University** is the most popular and rewarding spot for birding in Morgantown. The name honors Dr. Earl L. Core, distinguished WVU botanist who provided the inspiration and direction leading to the establishment and development of this wooded tract. The 50 acre Arboretum is situated between Monongahela Boulevard and the Monongahela River, and presents a mix of grassy lawn, open field, wooded hillside, and low, moist bottomland.

Core Arboretum is located approximately two miles west of downtown Morgantown, on Monongahela Boulevard. From north or south on I–79 take exit 155 and go approximately two miles east on US 19/SR 7. The Core Arboretum will be on your right, immediately past WVU Coliseum and its large parking lot. If arriving from the east on I–68, take exit 4 and proceed northwest on SR 7 through Morgantown to the Arboretum.

Walking from the Arboretum's entrance down to the river 200 feet below, a variety of habitats with several hundred species of native West Virginia trees, shrubs, and plants can be seen. These habitats and the Arboretum's location on a north–flowing river make it an excellent place to observe many species of birds year round.

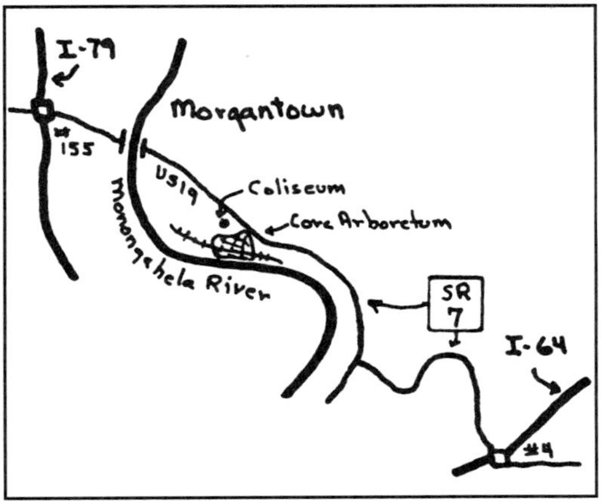

Three miles of well–maintained wooded trails and the newly developing rail trail that cuts through the Arboretum enable birders to take a short stroll or a extended walk. Open from dawn until dusk every day of the year, this wonderful outdoor laboratory is used by university and community folks alike. There is a small gravel parking area at the entrance, and no admission fee is charged. Trail maps and bird checklists are available free of charge at the trailhead just below the parking area.

Many local birders come here during spring migration to see as many as three dozen species of warblers as well as many other neotropical migrants. Nearly 200 species of birds have been observed at the Arboretum since it was opened to the public in 1954. Spring visitors are also treated to a dazzling display of wildflowers along the many walkways.

Spring early–morning bird walks are offered to the public free of charge during the months of April and May on Tuesdays and Fridays. Birders assemble at 7:15 a.m. in the parking area and spend about two hours leisurely birding their way down to the river and back. No reservations are required, and all levels of experience are welcomed. For more information, call Sally Stebbins at (304) 292–2128.

Coopers Rock State Forest

LeJay Graffious

Coopers Rock State Forest is a 12,000 acre mixed oak forest. The elevation of the park begins at 900 feet and climbs to 2610 feet above sea level. Coopers Rock State Forest is divided by I–68 and can be reached at exit 15 east of Morgantown or west of Bruceton Mills. Many well maintained trails snake through the forest. The managed Appalachian forest is the summer home for many wood warblers and other wildlife. Other attractions at Coopers Rock include **The Overlook, Henry Clay Iron Furnace, WVU demonstration area**, hunting, fishing and mountain bike trails. The Civilian Conservation Core developed The Overlook to peer into the 1200 foot deep view into the **Cheat River Canyon**, view the **Allegheny**

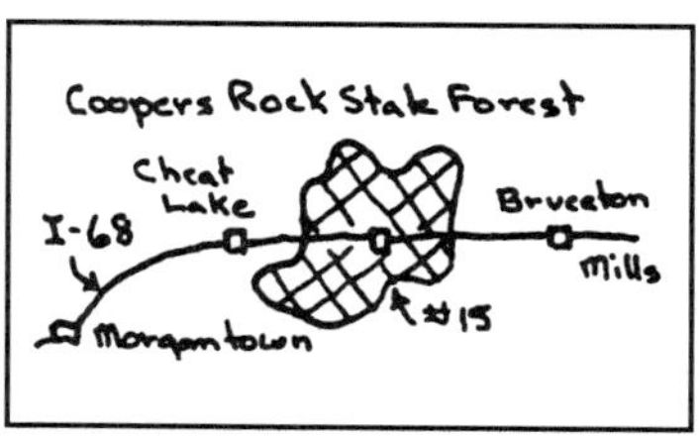

Highlands to the west and the plateau to the east. At the Overlook watch for raptors, ravens, and Turkey Vultures riding the thermals along the ridges. The Henry Clay Iron Furnace built in 1834 to purify the iron ore deposits in the area supported a community of about 500 people. At the Iron Furnace look for the Louisiana Waterthrush, Acadian Flycatcher, Wood Thrush and Black–throated Green Warbler.

One of the most unique birding spots is the **Virgin Hemlock Trail**. This climax forest trail is a beautiful 1.2 mile walk along a cool mountain stream with towering hemlocks and oaks. It is a great place for wood warblers from mid–May to the end of June. Along the trail watch and listen for Northern Parula, Black–and-white, Hooded, Black–throated Blue, Black–throated Green and Blackburnian warblers. Other species to be alert for are Ruffed Grouse, Wild Turkey, Solitary and Red–eyed vireo, Winter Wren, Scarlet Tanager, Wood Thrush, Blue–gray Gnatcatcher and Acadian Flycatcher.

The best bird watching area is at the **trails** that start at the **Archery Range**. These cross more diverse habitats that include meadows, shrubs and woodlands. In the trees surrounding the parking lot watch for American Redstart. Other birds found on these trails include Yellow–billed and Black–billed cuckoos, Pileated, Hairy, and Downy woodpeckers, Brown Creeper, Blue–gray Gnatcatcher, Chestnut–sided Warbler, Rose–breasted Grosbeak, Eastern Towhee and Broad–winged Hawk.

Other permanent avian residents are Coopers Hawk, Barred Owl, Blue Jay, Common Crow, Black–capped Chickadee, Carolina Wren, Tufted Titmouse, Cardinal, Song Sparrow and White–breasted Nuthatch. Other summer warblers found at Coopers Rock are Worm–eating, Cerulean, Prairie, Kentucky and Canada.

People who enjoy studying flora and fauna will find a wide variety. Some unusual species are Eastern Woodrat, Green Salamander, Flat–spired Three–toothed Land Snail, and mountain spleenwort.

More information may be obtained by writing: Coopers Rock Foundation, Inc., P.O. Box 505, Morgantown, WV, 26505–0505.

Cranesville Swamp

Norma Venable

Cranesville Swamp is a relict wetland complex supporting northern plants and animals at a southern latitude. Currently, The Nature Conservancy (TNC) owns over 300 acres in the **Cranesville Swamp Nature Sanctuary**. A boardwalk provides access to the mountain bog habitat. Visitors are welcome. Plants, wildlife, and the wetlands are protected.

Cranesville Swamp is located in West Virginia's Preston and Maryland's Garrett counties, nine miles north of Terra Alta, via CR 42 & 47. Roads in the Sanctuary are gravelled or dirt. From Terra Alta, proceed north towards Cranesville on SR 47 (Oak Grove Road) and turn right at the stone church onto Burnside Camp Road (CR 49). Proceed about a mile and turn left on the county road and follow TNC signs to the parking area. From Maryland, leave Cranesville on CR 47 and proceed east to the White–Rock Cranesville Road. Turn right and proceed south to the Methodist Church in Lake Ford. Turn right and follow TNC signs.

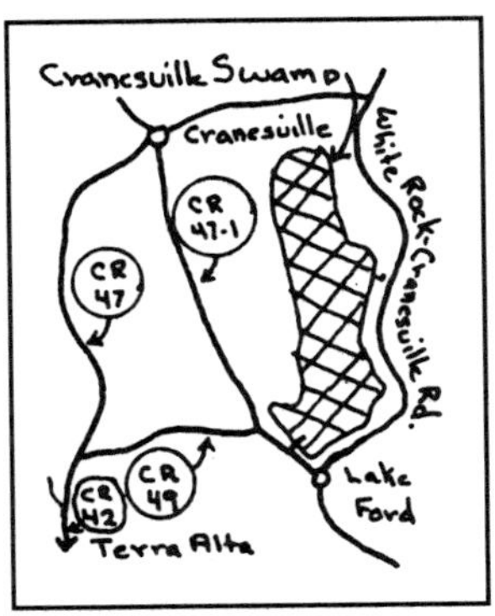

Visitors arriving at the Sanctuary park in the designated area and proceed along the road through the red pine woods until reaching the bog and the **boardwalk**, about a 10 minute walk.

Most of the walking to and around the Swamp is level. A leisurely trip to the Swamp and the boardwalk could be done in one to two hours. Mosquitoes and insects can be expected, especially on hot summer days. Fall and winter temperatures are often colder than surrounding areas.

Cranesville Swamp's unique aspect and scenery are readily apparent and you will better enjoy this fascinating place and have a better idea of what to look for if you know some background on the Swamp and how it was formed.

The history of the Swamp goes back thousands of years to the ice ages. Glaciers did not reach as far south as the present–day swamp, but the area was cold and conditions resembled those in Canada today. The region's plants and animals were northern species. In time, the climate gradually warmed, and the southern deciduous forest gradually reoccupied its former habitat. However, some glacial age northern plants and animals remained in cool places such as Cranesville Swamp, including such plants as goldthread, growing as far north as Alaska, Canada mayflower, growing as far north as Labrador, sundew, reaching Greenland, and the deciduous conifer eastern larch, growing today in Alaska. The eastern larch is an ice–age relict reaching its southernmost known location at Cranesville Swamp.

Cranesville Swamp's elevation of 2,560 feet keeps the swamp cool, with a short growing season, and an average 50 inches of rain annually preserves the boggy conditions required by much of its vegetation. Snowfall averages 160 inches and snow can occur even in summer. Poor drainage results in standing surface water which is acid because of decaying vegetation such as sphagnum moss. Thus, cold, wet, and acid conditions exist at the Swamp, influencing the plants and animals occurring here.

Cranesville Swamp was at one time logged, and this extensive cutting let in more light resulting in warming conditions and the growth of more southern plants.

Visitors walking along the boardwalk can easily view bog plants such as sundew, sphagnum moss, hair–cap moss, cottongrass, sedges, skunk cabbage, large cranberry, and cinnamon fern. The shrub thicket plant community includes Black chokeberry, winterberry, blueberry, rhododendron, mountain laurel, and viburnum.

A small pond at the beginning of the boardwalk is a good place to look for dragonflies, frogs, salamanders, and aquatic insects. Also, the boardwalk forms a square and inside this viewing space you should be able to find the dime–sized Sundew plants. At the end of the boardwalk you can see the Eastern Larch. The area around the boardwalk is open and birders should scan the sky and the surrounding woods for woodpeckers, sparrows, flycatchers, warblers, and raptors.

Birdlife in the swamp is varied, depending on season

and weather. High elevation species include Dark–eyed Junco, Northern Waterthrush, Solitary Vireo, Canada Warbler, American Redstart, Black–throated Green and Black–throated Blue warblers. Other fairly common birds include Gray Catbird, Swamp Sparrow, Ruffed Grouse, Indigo Bunting, Downy and Red–bellied woodpeckers, Wild Turkey, Green Heron, Yellow–billed Cuckoo, and birds of prey including hawks, falcons, and owls. The rare Northern Saw–whet Owl may nest on the Maryland side of the swamp. March nights are the time to listen for these little owls.

The wooded area around the swamp provides cover for birds and other wildlife. Swamp Sparrows nest in the shrubs around the parking lot, and the woods are home to such plants as gaywings, wild sarsaparilla, dwarf cornel, trailing arbutus, blue vervain, cardinal flower, and turtlehead.

Cathedral State Park

Norma Venable

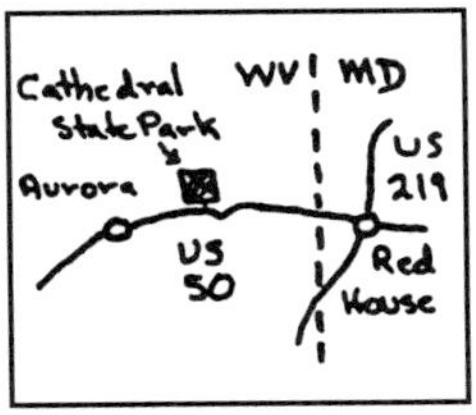

Cathedral State Park is a 133 acre stand with old growth hemlock trees, some over 400 years old. The park is located in Preston County, on US 50 near the Maryland and West Virginia state border. Cathedral State Park is one of the few stands of mixed old growth forest left in West Virginia. The trees are hemlock, along with yellow birch, red oak, and black cherry.

The park is maintained by the state as a natural area to permit biological processes to occur without human intervention. Trails are kept clear, but fallen trees are left to provide habitat for wildlife and to recycle nutrients to the soil. Because it is maintained as a natural area, no camping, hunting or fishing is permitted. There are picnic facilities, a covered pavilion and a play area for children. The trails are mostly level, and easy walking.

Birds associated with conifers namely the Black–throated Blue and Black–throated Green warblers as well as

Scarlet Tanager, Baltimore Oriole, Dark–eyed Junco, Rose–breasted Grosbeak, Magnolia and Parula warblers, Solitary Vireo, Wood and Hermit thrushes, and Barred and Eastern Screech–Owls can be seen.

More information may be obtained by writing Superintendent, Cathedral State Park, Aurora, WV, 26707. The Superintendent lives in residence at the cabin near the parking lot and conducts tours through the Park.

Pleasant Creek Wildlife Management Area

Joey Herron and Kyle Bush

Pleasant Creek was created by the construction of a flood control dam on the Tygart's Valley River near **Grafton**, West Virginia, thus creating Tygart Lake. Pleasant Creek arises in Barbour County and practically forms the border between Taylor and Barbour counties and empties into Tygart Lake.

When leaving Philippi, via the covered bridge, turn north on US 250/119 and travel just beyond the Taylor/Barbour county line turning right onto Pleasant Creek Road (CR 10). The wetlands are on the right side of the road with the best portion being 0.8 miles from the entrance. These wetlands were created by beavers damming Pleasant Creek. Continue to follow the road 2.4 miles and arrive at the **Doe Run Impoundment**, which was created by the Corps of Engineers for a breeding area for fish in Tygart Lake. Continue 0.2 mile further to the end of the road and the boat ramp in **Tygart Lake**. Migratory waterfowl may be found in the spring and fall, depending on the weather conditions and wind direction.

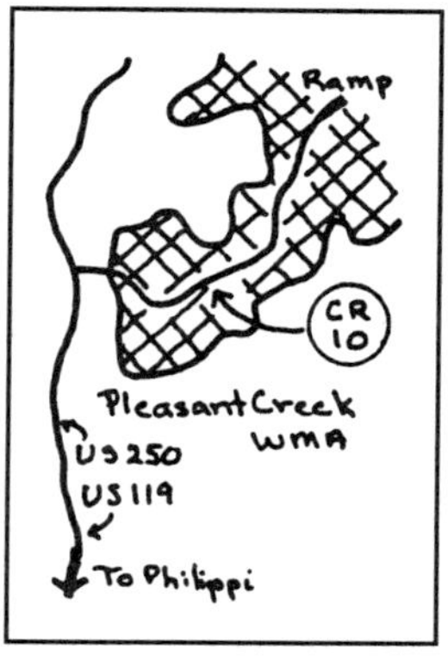

The following migratory waterfowl can be seen at Pleasant Creek and Doe Run from the middle of September until late December.

Common Loon
Pied–billed Grebe

Horned Grebe
Green Heron
Great Blue Heron
Tundra Swan

Snow Goose
Canada Goose
Wood Duck
Green–winged Teal
Blue–winged Teal
Mallard
Am. Black Duck
Northern Pintail
Gadwall
Northern Shoveler
American Wigeon
Canvasback
Redhead
Lesser Scaup
Bufflehead
Ring–necked Duck
Common Goldeneye
Ruddy Duck
Hooded Merganser
Common Merganser
Double–crested Cormorant

Audra State Park

Joey Herron and Kyle Bush

Audra State Park is located approximately 10 miles from Buckhannon, West Virginia, on the Upshur/Barbour County line. The park is split by the **Middle Fork River**, with a natural area on each side. There are three picnic areas, a swimming area in the river, a camping area with 62 sites and a hiking trail.

From Buckhannon, take US 33 east and turn left on CR 3. Proceed about 10 miles to the park. From Philippi, drive south on US 250 to Belington and turn right just south of town onto CR 11 which takes you to the park. The birds that can be found in the park are Broad–winged Hawk,

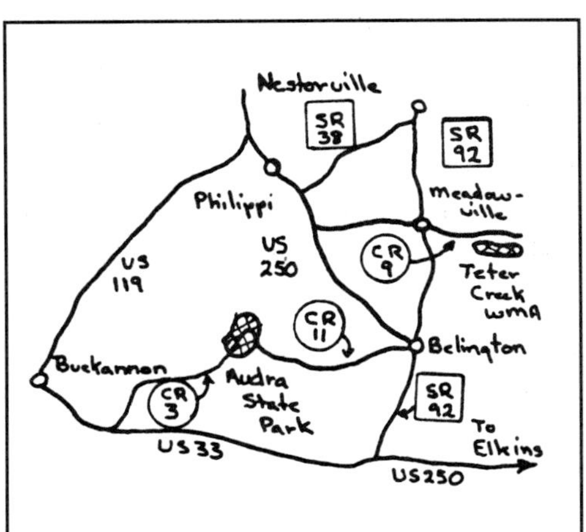

American Kestrel, Wild Turkey, Killdeer, Mourning Dove, Black-billed Cuckoo, Chimney Swift, Downy, Pileated and Red–bellied woodpeckers, Northern Flicker, Eastern Wood Pewee, Eastern Phoebe, Barn Swallow, Blue Jay, Great–crested Flycatcher, Tufted Titmouse, House Wren, Wood Thrush, American Robin, Mockingbird, Gray Catbird, Brown Thrasher, Red–eyed and Solitary vireos, Yellow, Hooded, and Chestnut–sided warbler, Common Yellowthroat, Ovenbird, Louisiana Waterthrush, Scarlet Tanager, Northern Cardinal, Indigo Bunting, Eastern Towhee, Chipping, Field and Song sparrows, Red–winged Blackbird, Eastern Meadowlark, Common Grackle Orchard and Baltimore oriole, Brown–headed Cowbird, House Finch, House Sparrow, American Goldfinch, and Starling.

A unique plant found by the river's edge is Barbara's buttons.

Teter Creek Wildlife Management Area

Joey Herron and Kyle Bush

Teter Lake is a public fishing area. Birding for waterfowl is like opening a Christmas package, you never know what you might find. The weather and the direction of the winds determine what will be there. If the weather is good, there are usually too many fishermen on the lake.

Teter Lake was created by the Department of Natural Resources of West Virginia as a fishing lake for stocked trout, large mouthed bass, and channel catfish. The lake is formed behind an earthen dam across Teter Creek at Kirt, West Virginia, which created a lake of 38.5 acres.

To reach Teter Lake from Philippi, (see Audra State Park map), go south on US 250 for two miles to SR 38 which goes to Nestorville. It is seven miles to Nestorville where SR 38 and 92 intersect. Turn south on SR 92 and go four miles to the Teter Lake sign in Meadowville. Take the Mid–way and Kirt Road (CR 9) four miles to the lake. Teter Lake is in the southeastern section of Barbour County at the foot of Laurel Mountain.

Here is a list of migratory waterfowl you may see at the lake:

Common Loon–Arrive late Oct.
Pied–billed Grebe–Arrive Oct.
Horned Grebe–Arrive Oct.
Great Blue Heron
Green Heron
Great Egret–Spring
Tundra Swan–Nov.
Snow Goose
Canada Goose–Breeding
Wood Duck–Oct. and March
Mallard–Breeding
Northern Pintail
Blue–winged Teal–March
Green–winged Teal–March
Northern Shoveler
Gadwall–Oct. and Nov.
Amer. Wigeon–Oct. and Nov.
Canvasback
Redhead Duck
Ring–necked Duck–Common
Lesser Scaup–Oct. through Dec.
Black Scoter
White–winged Scoter
Common Goldeneye–Oct./Dec.
Bufflehead–East Winds/Oct.–Dec.
Oldsquaw–Dec.
Hooded Merganser–Nov.–Dec.
Ruddy Duck–Oct. and Dec.
Red–br. Merganser–Nov.–Dec.
Greater Yellowlegs
Spotted Sandpiper
Ring–billed Gull
Herring Gull
Bonaparte's Gull

Sleepy Creek- Male Cerulean Warblers can be located throughout the breeding season, singing in the mid to upper canopy along the ridgetop and upper eastern slope of Sleepy Creek Mountain, both north and south of "The Locks".

North Bend Rail Trail- The Cerulean Warbler can be found along the rail trail in several places. The first is from the intersection of Way Station Road and the rail trail. In the first few miles going west, there are several areas where Cerulean Warblers are found on both sides of the trail. The second area is accessed by CR 5, Bear Run Road, off US 50. Go east on the trail about a mile and between the first tunnel near former town of Cornwallis and the second tunnel, there are good Cerulean Warbler areas on the north side of the trail.

Coonskin Park

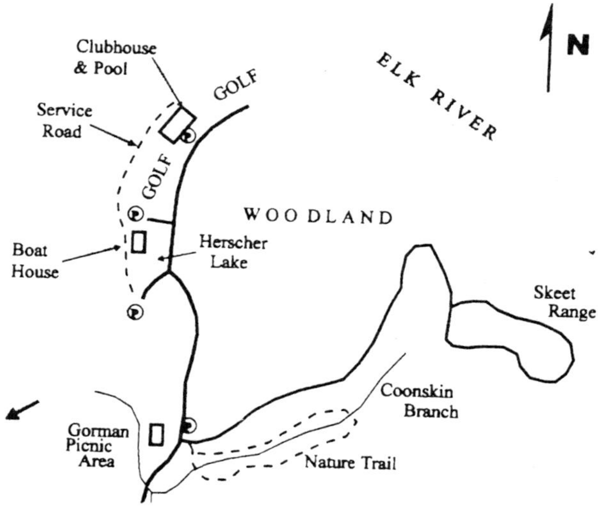

Part Three

Other Information

Noteworthy Plants
Reptiles and Amphibians
Information Centers
Radio and Police
Organizations
Contact Persons
Publications
Bird Checklist

Noteworthy Plants of West Virginia

Reba and Gene Hutton

The following plants represent a small part of the flora of WV. They are listed in phylogenetic order according to Core's "Flora of WV", second edition. The sections are shown on the physiographic map on page 11.

Western Hill Section I

May/June	Green Dragon	*Arisaema dracontium*
May/June	Sm.Yl. Lady Slipper	*Cypripedium parvifolorum*
Apr/June	Lg.Yl. Lady Slipper	*Cypripedium pubescens*
June/Aug	Roundedleaf Catchfly	*Silene rotundifolia*
Apr/May	Goldenseal	*Hydrastis canadensis*
Apr/May	Twinleaf	*Jeffersonia diphylla*
Mar/Apr	Pawpaw	*Asimina triloba*
May/June	Celandine Poppys	*Stylophorum diphyllum*
July/Aug	Viginia Mallow	*Sida hermaphrodita*
June/July	Galax	*Galax aphylla*
June/Aug	Butterfly Weed	*Asclepias tuberosa*
Apr/June	Miami Mist	*Phacelia purshii*
Mar/Apr	Virginia Bluebells	*Mertensia virginica*
June/Sept	Dragon Head	*Dracocephalum virginianum*
May/June	Guyandotte Beauty	*Synandra hispidula*
Apr/June	Blue-eyed Mary	*Collinsia verna*
June/Sept	Culver's Root	*Veronicastrum virginicum*
Apr/June	Crossvine	*Bignonia capreolata*
Apr/June	Elecampana	*Inula helenium*
Aug/Sept	Golden Knees	*Chrysogonum virginianum*

Allegheny Mountain and Upland Section II

May/June	Fraser's Sedge	*Carex fraseri*
Apr/May	Golden Club	*Orontium aquaticum*
May/June	Tofieldia	*Tofieldia glutinosa*
May/June	Devil's Bit	*Chamaelirium luteum*
Aug/Sept	Featherbells	*Stenanthium gramineum*
Apr/June	Ramps	*Allium tricoccum*

Mar/Apr	Snow Trillium	*Trillium nivale*
May/June	Painted Trillium	*Trillium undulatum*
June/Aug	Lg. Purple Fringed Orchid	*Plantanthera fimbriata*
July/Aug	Yl. Fringed Orchid	*Plantanthera ciliaris*
Apr/June	Marsh Marigold	*Caltha palustris*
May/Aug	Goldthread	*Coptis groenlandica*
June-July	White Monkshood	*Aconitum reclinatum*
May/Sept	Wild Bleeding Heart	*Dicentra eximia*
May/Aug	Three-toothed Cinquefoil	*Potentilla tridentata*
May/June	Gay Wings	*Polygala paucifolia*
May/July	Poison Sumac	*Rhus vernix*
May/June	Bog Rosemary	*Andromeda glaucophylla*
Apr/June	Buckbean	*Menyanthes trifoliata*
May/June	Jacob's Ladder	*Polemonium van-bruntiae*
June/Aug	Barbara's Buttons	*Marshallia grandiflora*

Eastern Ridge and Valley Section III

Apr/Aug	Holy/Sweet Grass	*Hierochloe odorata*
June/July	Turkey Beard	*Xerophyllum asphodeloides*
Apr/May	Sweet Fern	*Comptonia peregina*
July/Aug	Yellow Buckwheat	*Eriogonum alleni*
May/Aug	Fameflower	*Talinum teretifolium*
July/Sept	Silvery Nailwort	*Paronychia argyrocoma*
May/June	Shale Leatherflower	*Clematis albicoma*
Aug/Oct	Grass of Parnassus	*Parnassia asarifolia*
Apr/May	Kates Mountain Clover	*Trifolium virginicum*
Mar/June	Bent Milkvetch	*Astragalus distortus*
Apr/May	Toothache Tree	*Zanthoxylum americanum*
Mar/Apr	Fragrant Sumac	*Rhus aromatica*
May/Sept	Canby's Mountain Lover	*Pachistima canbyi*
May/June	False Heather	*Hudsonia tomentosa*
Apr/June	Bird-foot Violet	*Viola pedata*
Apr/May	Silverbell	*Halesia carolina*

Apr/June	Orange Puccoon	Lithospermum canescens
June/Aug	Twinflower	Linnaea borealis
Aug/Oct	Gay Feathers	Liatris spicata
Aug/Sept	Golden Aster	Heterotheca mariana
Aug/Oct	Stiff Aster	Aster linarii-folius

Other Invertebrates

Jane Whitaker

Amphibians–The amphibian populations of West Virginia vary dramatically due to changes in altitude from the Ohio Valley (at approximately 600 ft. above sea level) to the high ridges of the Appalachians in the eastern counties (to over 4800 ft.).

The Hellbender, *Cryptobrachus alleganiensis*, a relic aquatic salamander that can grow up to 20 inches in length is found throughout the Ohio River drainage of the state.

Frogs and toads are found throughout West Virginia, however some can be found only in the eastern counties. All species can be identified by sound. The use of a good recording of their calls, particularly in the spring and early summer, can enhance one's ability to identify them.

Lungless salamanders, *Plethondontidae*, are common throughout the state. Most of the genus *Desmognathus* are found near streams or in very damp areas. Those of the genus *Plethodon* are primarily woodland species. Some of these are endemic to certain geographic areas of the state. In some cases a species can be found on only one or two mountains. Amphibians and Reptiles In West Virginia by Green and Pauley, Univ. of Pitt. Press. 1987, is highly recom--mended if one is inclined to observe, identity, and become familiar with the amphibians and reptiles of the state.

Reptiles–A number of the turtle species found in West Virginia are often observed in isolated communities. There are no large wetland areas and there is only one natural pond in the state. The largest variety of these turtle populations is found in areas near the Ohio River in the west, the Potomac River in the Eastern Panhandle, and the extreme southern part of the state. A few species such as the Snapping Turtle and the terrestrial Box Turtle are found throughout the state.

There are two poisonous snakes in the state. Copperheads are found throughout West Virginia. However, the Timber Rattlesnake is rarely found at lower elevations near the Ohio River. There are over 20 species of snakes and 5 species of lizards within the borders of the state. As with turtles, some are more widely distributed that others.

Many of the native reptiles are now being collected as exotics to be sold domestically and overseas. Collecting reptiles and some amphibians without a valid collecting permit is illegal in many states in order to preserve the species. Snakes are also frequently killed indiscriminately by individuals who do not understand their ecological importance.

Information Centers

I–64 Cabell County, eastbound lanes near WV 10–Huntington interchange

I–64 Greenbrier County, westbound lanes east of White Sulfur Springs

I–70 Ohio County, westbound lanes 1/2 mile west of state line

I–77 Wood County, southbound lanes south of Mineral Wells interchange

I–77 Kanawha County, northbound lanes at mile post 72, Morton

I–77 Mercer County, northbound lanes at mile post 18, Bluestone

I–79 Monongahela County, southbound lanes north of Morgantown

I–81 Berkeley County, northbound lanes two miles north of Virginia state line

I–81 Berkeley County, southbound lanes one mile south of the Maryland state line

US 340 Harpers Ferry

Radio and Police

West Virginia Public Radio

Charleston	88.5	WVPN
Elkins	88.5	
Buckhannon	88.9	WVPW
Martinsburg	88.9	WVEP
Huntington	89.9	WVWV
Wheeling	89.9	WVNP
Parkersburg	90.3	WVPG
Morgantown	90.9	WVPM
Beckley	91.7	WVPB
Clarksburg	107.3	

West Virginia State Police
911

West Virginia Organizations

The following list is composed of a few ecologically-minded organizations that may be helpful to your visit to West Virginia.

The Brooks Bird Club of West Virginia
P.O. Box 4077
Wheeling, WV, 26003

Handlan Chapter of the BBC
Jim Triplett 1-304-744-4913

Bibbee Nature Club
Jim Phillips 1-304-466-1275

Tri-State Audubon (Huntington)
Mike Forman 1-304-453-2301

Vandalia Audubon (Charleston)
Pat Barker 1-304-345-8143

Three Rivers Raptor Center
1-800-721-5252

WV Native Plant Society
John Northeimer 1-304-259-5377

WV Rails-to-Trails Council
P.O. Box 85
Nitro, W.V 25143
1-304-722-6558

WV Scenic Trails Association
Roger Hardway (Charleston) 1-304-744-6979
Vicky Shears (Morgantown) 1-304-285-4083

WV Department of Natural Resources and Natural Heritage Program
P. J. Harmon 1-304-637-0245

Volunteers

The following list is made up of those who are responsible for creating the bird guide. Those with accompanying phone numbers are willing to receive calls and answer questions about the current birding and botanizing conditions in their region.

Ken Anderson	304-344-1446
Wendell Argabrite	304-736-5747
Ralph Bell	412-883-4505
George Breiding	304-296-4630
Dana Buckelew	404-892-4378
Jay Buckelew	304-829-4392
Jim and Beth Bullard	304-799-7345
Eleanor Bush	304-457-3460

Ronald Canterbury	304-384-5214
Virginia Cronenberger	304-628-3382
Bob Dean	304-274-1161
Greg Eddy	304-242-4774
Cindy Ellis	304-586-4135
Dick and Jeanette Esker	304-863-8765
Tom and Dawn Fox	304-354-7686
LeJay and Helen Ann Graffious	304-379-4492
Emily Grafton	304-292-0229
Mike Griffith	304-522-4674
Hullet Good	304-346-6990
George Hall	304-594-2521
Joey Herron	304-363-5553
Gene and Reba Hutton	304-636-1598
Marjorie Keatley	304-523-7512
Nevada Laitsch	216-385-0582
Josh Lipton	304-647-9931
Karen McClure	
Grady and Anne McRae	304-327-7556
Jim Miller	304-536-2144
Bill Murray	304-564-5071
John Northeimer	
Diane Packett	765-463-9231
Jim Phillips	304-466-1275
Mary Moore Rieffenberger	304-636-4559
Bob Rine	304-723-2205
Carl Slater	614-635-9246
Clete Smith	412-561-7267
J. Lawrence Smith	304-727-7103
Sally Stebbins	304-599-7015
Bill Thompson	614-473-1141
Arline Thorn	304-722-6346
James F. Triplett	
Norma Venable	304-328-5128
Jane and Matthew Whitaker	202-966-6218
Doug Wood	304-466-2724
Julie Zickefoose	614-473-1141

Publications

Birds of West Virginia, George Hall, Carnegie Museum of Natural History, Pittsburgh

The West Virginia Breeding Bird Atlas, Albert R. Buckelew, Jr. and George Hall, University of Pittsburgh Press

A Guide to the Birds of the New River Gorge Area, Written and compiled by Gary Worthington, James D. Phillips, and Don Kodak, Department of Natural Resources, West Virginia

Birds of the Lower Ohio River Valley in West Virginia, Lloyd Kiff, Thomas D. Igou, Harry E. Slack III, and Leon Wilson, S special Publication #1 of the Brooks Bird Club.

Birds of the West Virginia Panhandle, Albert R. Buckelew, Jr., Published by The Brooks Bird Club

Endangered and Threatened Species in West Virginia, Albert R. Buckelew Jr., A Special Publication #2 by The Brooks Bird Club

Flora of West Virginia, P.D. Strausbaugh and Earl L. Core, Seneca Books, Inc., Grantsville, West Virginia

Hiking the Mountain State, The Trails of West Virginia, Allen de Hart, Appalachian Mountain Club, Boston, Massachusetts

Monongahela National Forest Hiking Guide, Bruce Sundquist and Allen de Hart for the West Virginia Highlands Conservancy, Charleston, West Virginia

Rail Trails Along the Greenbrier , Jim Hedson, Quarrie Press, Charleston, West Virginia

Check List of the Birds of West Virginia
West Virginia Nongame Wildlife
and
Natural Heritage Program

It is with appreciation that the following draft of the new Check List compiled and published jointly by the West Virginia Nongame Wildlife and Natural Heritage Program of The Department of Natural Resources of The State of West Virginia and the Brooks Bird Club of West Virginia is presented in this guide. The codes are as follows:

Habitat Codes

E=Forest/Field Edges
F=Fields and Clearings
G=Generalist, Variety of Habitats
H=Hardwood Forests
M=Mountain Cliffs
P=Pine Forests
R=Towns and Residential Areas
S=High Elevation Spruce Forests
T=Thickets, Heaths
W=Wetlands, Streams, Open Waters

Status Codes

E=Exotic, Introduced Species
M=Spring and Fall Migrant
P=Permanent, Occurs in WV year-round
S=Summer Resident
V=Visitant, appears in WV irregularly
W=Winter Resident

Abundance Codes

C=Commonly Encountered in WV
O=Occasionally Encountered in WV
R=Rarely Encountered in WV

(N)=Neotropical Migrant: A species most of whose migrant population winters is Latin America or the Caribbean

*Species with fewer than five accepted records in West Virginia. Send complete documentation of any observation of these or any species not on this list to West Virginia Records Committee, West Virginia Nongame Wildlife Program, P.O. Box 67, Elkins, WV 26241

**Send nesting records to West Virginia Nongame Program, P.O. Box 67, Elkins, WV, 26241

Check List of Birds of West Virginia

	Habitat	Status	Abundance
Loons			
___Red–throated Loon	W	M	R
___Common Loon	W	M	O
Grebes			
___Pied–billed Grebe	W	M	O
___Horned Grebe	W	M	O
___Red–necked Grebe	W	M	R
Pelican			
___American White Pelican			
Cormorants			
___Double–Crested Cormorant	W	M	R
___ Great Cormorant*			
Herons & Bitterns			
___American Bittern**	W	S,M	O
___Least Bittern (N)	W	S,M	O
___Great Blue Heron	W	P	O
___Great Egret	W	S	O
___Snowy Egret	W	V	R
___Little Blue Heron (N)	W	V	R
___Cattle Egret**	F,W	V,E	R
___Green Heron	W	S	C
___Black–crowned Night–Heron**	W	M	O
___Yellow–crowned Night–Heron**	W	M	R
Ibis			
___ White Ibis*			
___ Glossy Ibis*			
Storks			
___ Wood Stork			
Vultures			
___Black Vulture	G	S	O
___Turkey Vulture	G	S	C
Waterfowl			
___ Black-bellied Whistling-Duck*			
___ Greater White-fronted Goose			
___Snow Goose	W	M	R
___ Ross's Goose*			
___Canada Goose	W,F	P	C
___ Brant			
___ Barnacle Goose* (pending acceptance by the Records Com.)			

___Mute Swan**	W	V,E	O
___ Trumpeter Swan*			
___Tundra Swan	W	M	O
___Wood Duck	W	S	C
___Gadwall	W	M	O
___ Eurasian Wigeon*			
___American Wigeon	W	M	C
___American Black Duck	W	P	C
___Mallard	W	P	C
___Blue–winged Teal	W	S,M	O
___Northern Shoveler	W	M	O
___Northern Pintail (N)	W	M	O
___Green–winged Teal**	W	S,M	O
___Canvasback	W	M	O
___Redhead	W	M	C
___Ring–necked Duck	W	M	C
___Greater Scaup	W	M	R
___Lesser Scaup	W	M	C
___King Eider*			
___Surf Scoter	W	M	R
___White–winged Scoter	W	M	O
___Black Scoter	W	M	R
___ Oldsquaw	W	M,W	R
___Bufflehead	W	M	O
___Common Goldeneye	W	M	C
___Hooded Merganser	W	S,M	O
___Common Merganser	W	M	R
___Red–breasted Merganser	W	M	O
___Ruddy Duck	W	M	O
Eagles & Hawks			
___Osprey (N)**	W	S,M	R
___ Swallow-tailed Kite*			
___Bald Eagle**	H,W	S	R
___Northern Harrier**	W	S,M	O
___Sharp–shinned Hawk	H,S	P	O
___Cooper's Hawk	H,E	P	O
___Northern Goshawk**	H,S	P	R
___Red–shouldered Hawk	H	S	O
___Broad–winged Hawk (N)	H	S,M	O
___ Swainson's Hawk*			
___Red–tailed Hawk	E,F	P	C

___Rough–legged Hawk	E,F	M,W	R
___Golden Eagle	M,G	M	R
Falcons			
___American Kestrel	E,F	P	C
___Merlin (N)	M,S	M	R
___Peregrine Falcon (N)**	M	S,M	R
Gamebirds			
___Ring–necked Pheasant	F,E	P,E	O
___Ruffed Grouse	H	P	C
___Wild Turkey	H,F	P	C
___Northern Bobwhite	F	P	O
Rails, Gallinules & Coots			
___ Yellow Rail*			
___King Rail	W	M	R
___Virginia Rail**	W	S,M	R
___Sora	W	S,M	R
___ Purple Gallinule			
___Common Moorhen(N)**	W	M	R
___American Coot**	W	M	O
Crane			
___ Sandhill Crane			
Plovers			
___Black–bellied Plover (N)	W	M	R
___ American Golden-Plover (N)	W	M	R
___Semipalmated Plover	W	M	R
___ Piping Plover*			
___Killdeer	F,W	P	C
Stilts & Avocets			
___ Black-necked Stilt*			
___ American Avocet			
Sandpipers			
___Greater Yellowlegs	W	M	O
___Lesser Yellowlegs	W	M	O
___Solitary Sandpiper	W	M	R
___ Willet			
___Spotted Sandpiper	W	S,M	O
___Upland Sandpiper (N)**	F	S,M	R
___ Whimbrel*			
___ Hudsonian Godwit*			
___Marbled Godwit (N)*	W	M	R
___Ruddy Turnstone (N)	W	M	R
___ Sanderling			

___Semipalmated Sandpiper (N)	W	M	C
___Western Sandpiper (N)	W	M	R
___Least Sandpiper (N)	W	M	O
___White–rumped Sandpiper (N)	W	M	R
___Baird's Sandpiper (N)	W	M	R
___Pectoral Sandpiper (N)	W	M	O
___Dunlin	W	M	R
___Stilt Sandpiper (N)	W	M	R
___Short–billed Dowitcher (N)	W	M	R
___Long–billed Dowitcher*	W	M	R
___Common Snipe**	W	S,M	O
___American Woodcock	W,T	S,M	O
Phalaropes			
___Wilson's Phalarope (N)	W	V	R
___Red–necked Phalarope (N)	W	V	R
___ Red Phalarope*			
Gulls & Terns			
___ Parasitic Jaeger*			
___Laughing Gull	W	M	R
___Franklin's Gull (N)*	W	M	R
___ Little Gull* (pending acceptance by Records Com)			
___Bonaparte's Gull	W	M	C
___Ring–billed Gull	W	M	C
___Herring Gull	W	M,W	O
___ Sabine's Gull*			
___ Black-legged Kittiwake*			
___Caspian Tern	W	M	O
___Common Tern (N)	W	M	R
___Forster's Tern (N)	W	M	R
___ Leat Tern			
___ Sooty Tern*			
___Black Tern (N)	W	M	R
Doves			
___Rock Dove	M,R	P,E	C
___Mourning Dove	F,R	P	C
Cuckoos			
___Black–billed Cuckoo (N)	H	S,M	O
___Yellow–billed Cuckoo (N)	H,T,E	S,M	C
Owls			
___Barn Owl**	E,F	P	R
___Eastern Screech–Owl	H,E	P	C
___Great Horned Owl	H,E	P	C

___Snowy Owl	F	W	R
___Barred Owl	H	P	C
___Long–eared Owl**	P	P	R
___Short–eared Owl	F,W	M	R
___Northern Saw–whet Owl	S,P,H	P	R
Goatsuckers			
___Common Nighthawk (N)	F,R,M	S,M	O
___Chuck-will's-widow (N)**	H,P	S	R
___Whip–poor–will (N)	H,E	S,M	O
Swifts			
___Chimney Swift (N)	R,F	S,M	C
Hummingbirds			
___Ruby–thr. Hummingbird (N)	E,R	S,M	C
___ Rufous Hummingbird*			
Kingfishers			
___Belted Kingfisher (N)	W	P	C
Woodpeckers			
___Red–headed Woodpecker	H,E	P	R
___Red–bellied Woodpecker	H	P	C
___Yellow–bellied Sapsucker**	H,S	P	R
___Downy Woodpecker	H,E	P	C
___Hairy Woodpecker	P,H,S	P	C
___Northern Flicker	H,S	S,M	C
___Pileated Woodpecker	H	P	C
Flycatchers			
___Olive–sided Flycatcher (N)**	S	S	R
___Eastern Wood-Pewee (N)	H	S	C
___Yellow–bellied Flycatcher (N)**	S	S,M	R
___Acadian Flycatcher (N)	H	S	C
___Alder Flycatcher (N)	T,W	S	O
___Willow Flycatcher (N)	T,W	S	O
___Least Flycatcher (N)	H,E	S	O
___Eastern Phoebe (N)	E,M	S	C
___ Ash-throated Flycatcher*			
___Great Crested Flycatcher (N)	H	S	C
___ Western Kingbird			
___Eastern Kingbird (N)	E,F	S	C
Shrikes			
___Loggerhead Shrike	F,T	S	R
___ Northern Shrike*			
Vireos			
___White–eyed Vireo (N)	T,E	S	C

___Yellow–throated Vireo (N)	H	S	C
___Blue-headed Vireo (N)	H,S	S	O
___Warbling Vireo (N)	H	S	O
___Philadelphia Vireo (N)	H,E	M	R
___Red–eyed Vireo (N)	H	S	C
Jays & Crows			
___Blue Jay	H,R	P	C
___ Black-billed Magpie*			
___American Crow	H,F,E	P	C
___Fish Crow**	H,W	P	R
___Common Raven	H,S,M	P	O
Larks			
___Horned Lark	F	P	O
Swallows			
___Purple Martin (N)	R,F	M,S	C
___Tree Swallow (N)	H,E	M,S	C
___N. Rough–wg Swallow (N)	E,W,F	M,S	C
___Bank Swallow (N)	W	M	R
___Cliff Swallow (N)	F	S,M	O
___Barn Swallow (N)	F	S	C
Chickadees & Titmice			
___Carolina Chickadee	H,P	P	C
___Black–capped Chickadee	H,P,S	P	C
___ Boreal Chickadee*			
___ Tufted Titmouse	H,P	P	C
Nuthatches			
___Red–breasted Nuthatch	S,H,	P	O
___White–breasted Nuthatch	H	P	C
Creepers			
___Brown Creeper	H,S	P	O
Wrens			
___Carolina Wren	H,R	P	C
___Bewick's Wren**	F	S	R
___House Wren	F,T,R	S	C
___Winter Wren	S,H	P	O
___Sedge Wren**	F,W	S	R
___Marsh Wren**	W	S,M	R
Kinglets & Gnatcatchers			
___Golden–crowned Kinglet	S,H	P	O
___Ruby–crowned Kinglet	H	M,W	O
___Blue–gray Gnatcatcher (N)	H	S	C

Thrushes			
___Eastern Bluebird	F,E	P	C
___Veery (N)	H,S	S,M	O
___Gray–cheeked Thrush (N)	S	M	R
___Bicknell's Thrush*	M	M	R
___Swainson's Thrush (N)	S	S,M	O
___Hermit Thrush (N)	S,H	S,M	O
___Wood Thrush (N)	H	S	C
___American Robin	G	P	C
___ Varied Thrush*			
Mockingbirds & Thrashers			
___Gray Catbird (N)	H,T	S	C
___Northern Mockingbird	T,F	P	C
___Brown Thrasher	T,F	S	C
Starlings			
___European Starling	G	P,E	C
Pipits			
___American Pipit	W	M	O
Waxwings			
___ Bohemian Waxwing			
___Cedar Waxwing	F,T	P	C
Wood Warblers			
___Blue–winged Warbler (N)	H,T	S	C
___Golden–winged Warbler (N)	H,T	S	O
___Brewster's Hybrid Warbler (N)	H,T	S	R
___Lawrence's Hybrid Warbler (N)	H,T	S	R
___Tennessee Warbler (N)	G	M	C
___Orange–crowned Warbler (N)	H,S	M	R
___Nashville Warbler (N)**	E,W	S,M	O
___Northern Parula (N)	H	S	C
___Yellow Warbler (N)	E,T	S	C
___Chestnut–sided Warbler (N)	F,T	S,M	C
___Magnolia Warbler (N)	T,S	S,M	O
___Cape May Warbler (N)	G	M	O
___Black–throated Blue Warbler (N)	S,H	S,M	C
___Yellow–rumped Warbler (N)	S,T	M,S	O
___Black–throated Green Warbler (N)	H	S,M	C
___Blackburnian Warbler (N)	S,H	S,M	O
___Yellow–throated Warbler (N)	H	S	C
___Pine Warbler (N)	P	S	O
___Prairie Warbler (N)	H,T	S	C

___Palm Warbler (N)	F	M	R
___Bay–breasted Warbler (N)	G	M	O
__ Blackpoll Warbler (N)	G	M	O
___Cerulean Warbler (N)	H	S	C
___Black-and-white Warbler (N)	H	S,M	C
___American Redstart (N)	H	S,M	C
___Prothonotary Warbler (N)	W,H	S	R
___Worm–eating Warbler (N)	H	S	C
___Swainson's Warbler (N)	H,T	S	R
___Ovenbird (N)	H	S	C
___Northern Waterthrush (N)	W,S	S,M	O
___Louisiana Waterthrush (N)	H,W	S	C
___Kentucky Warbler (N)	H	S	C
___Connecticut Warbler (N)	T	M	R
___Mourning Warbler (N)	S,H,T	S	O
___Common Yellowthroat (N)	T,F	S	C
___Hooded Warbler (N)	H	S	C
___Wilson's Warbler (N)	G	M	O
___Canada Warbler (N)	H,S,T	S	O
___Yellow–breasted Chat (N)	T	S	C
Tanagers			
___Summer Tanager (N)	H	S	C
___Scarlet Tanager (N)	H	S	C
Towhees & Sparrows			
___ Green-tailed Towhee*			
___ Spotted Towhee*			
___ Eastern Towhee	H,F	P	C
___Bachman's Sparrow**	F	S	R
___American Tree Sparrow	F	W	O
___Chipping Sparrow	R,E,T	S	C
___ Clay-colored Sparrow*			
___Field Sparrow	E,F	S	C
___Vesper Sparrow	F	S,M	O
___Lark Sparrow**	F	S	R
___ Lark Bunting*			
___Savannah Sparrow	F	S	O
___Grasshopper sparrow	F	S	O
___Henslow's Sparrow**	F	S	R
___ Le Conte's Sparrow*			
__ Nelson's Sharp–tail Sparrow	F,T,W	M	R
___Fox Sparrow	F,T	W,M	O
___Song Sparrow	F,T,R	P	C

___Lincoln's Sparrow	T	M	R
___Swamp Sparrow	W	P	O
___White–throated Sparrow**	F,T,W	W,M,S	O
___Harris' Sparrow	E	W	R
___White–crowned Sparrow	P,H	W,M	O
___Dark–eyed Junco	S,H,E	P	C
___ Lapland Longspur			
___ Smith's Longspur*			
___ Snow Bunting			
Cardinals & Grosbeaks			
___Northern Cardinal	F,T,R	P	C
___Rose–breasted Grosbeak (N)	H	S	O
___ Black-headed Grosbeak			
___Blue Grosbeak (N)	F,T	S	O
___Indigo Bunting (N)	E	S	C
___Dickcissel (N)	F	S,M	R
Blackbirds & Orioles			
___Bobolink (N)	F	S,M	O
___Red–winged Blackbird	F,W	P	C
___Eastern Meadowlark	F	P	C
___ Western Meadowlark			
___ Yellow-headed Blackbird			
___Rusty Blackbird	T,W	M,W	O
___ Brewer's Blackbird			
___Common Grackle	F,R	S,M	C
___Brown–headed Cowbird	H	P	C
___Orchard Oriole (N)	F,H	S	O
___Baltimore Oriole (N)	F,R	S	C
Finches			
___Pine Grosbeak	H	W	R
___Purple Finch	S,H	P	O
___House Finch	R	P	C
___Red Crossbill**	S,P	W,S	O
___White–winged Crossbill	S	W	O
___Common Redpoll	T	V	O
___ Hoary Redpoll*			
___Pine Siskin	H,S	W,M,S	O
___American Goldfinch	E,F,R	P	C
___Evening Grosbeak	S,H,R	V	O
Weaver Finches			
___House Sparrow	R,F	P,E	C

Index